Even You Can Learn Statistics

Second Edition

A Guide for Everyone Who Has Ever Been Afraid of Statistics

David M. Levine, Ph.D.

David F. Stephan

Vice President, Publisher: Tim Moore
Associate Publisher and Director of Marketing: Amy Neidlinger
Executive Editor: Jim Boyd
Editorial Assistant: Myesha Graham
Operations Manager: Gina Kanouse
Senior Marketing Manager: Julie Phifer
Publicity Manager: Laura Czaja
Assistant Marketing Manager: Megan Colvin
Cover Designer: Alan Clements
Managing Editor: Kristy Hart
Project Editor: Anne Goebel
Copy Editor: Paula Lowell
Proofreader: Williams Woods Publishing
Interior Designer: Argosy
Compositor: Jake McFarland
Manufacturing Buyer: Dan Uhrig

FT Press offers excellent discounts on this book when ordered in quantity for
bulk purchases or special sales. For more information, please contact U.S.
Corporate and Government Sales, 1-800-382-3419, corpsales@pearsontech-
group.com. For sales outside the U.S., please contact International Sales at
international@pearson.com.

Second Printing January 2010

ISBN-10: 0-13-701059-1
ISBN-13: 978-0-13-701059-2

Pearson Education LTD.
Pearson Education Australia PTY, Limited.
Pearson Education Singapore, Pte. Ltd.
Pearson Education North Asia, Ltd.
Pearson Education Canada, Ltd.
Pearson Educación de Mexico, S.A. de C.V.
Pearson Education—Japan
Pearson Education Malaysia, Pte. Ltd.
Library of Congress Cataloging-in-Publication Data

Levine, David M., 1946-
Even you can learn statistics : a guide for everyone who has ever been afraid
of statistics / David M. Levine and David F. Stephan. – 2nd ed.
    p. cm.
ISBN 978-0-13-701059-2 (pbk. : alk. paper) 1. Statistics–Popular works.
I. Stephan, David. II. Title.
QA276.12.L485 2010
519.5–dc22
                                          2009020268

To our wives
Marilyn and Mary

To our children
Sharyn and Mark

And to our parents
In loving memory, Lee, Reuben, Ruth, and Francis

Table of Contents

Acknowledgments viii

About the Authors ix

Introduction   *The Even You Can Learn Statistics* Owners Manual xi

Chapter 1   Fundamentals of Statistics 1
   1.1 The First Three Words of Statistics 2
   1.2 The Fourth and Fifth Words 4
   1.3 The Branches of Statistics 5
   1.4 Sources of Data 6
   1.5 Sampling Concepts 7
   1.6 Sample Selection Methods 9

Chapter 2   Presenting Data in Charts and Tables 19
   2.1 Presenting Categorical Variables 19
   2.2 Presenting Numerical Variables 26
   2.3 Misusing Charts 32

Chapter 3   Descriptive Statistics 43
   3.1 Measures of Central Tendency 43
   3.2 Measures of Position 47
   3.3 Measures of Variation 51
   3.4 Shape of Distributions 57

Chapter 4   Probability 71
   4.1 Events 71
   4.2 More Definitions 72
   4.3 Some Rules of Probability 74
   4.4 Assigning Probabilities 77

Chapter 5   Probability Distributions 83
   5.1 Probability Distributions for Discrete Variables 83
   5.2 The Binomial and Poisson Probability Distributions 89
   5.3 Continuous Probability Distributions and the Normal Distribution ... 97
   5.4 The Normal Probability Plot 105

Chapter 6   Sampling Distributions and Confidence Intervals 119
   6.1 Sampling Distributions 119
   6.2 Sampling Error and Confidence Intervals 123

6.3 Confidence Interval Estimate for the Mean Using the t Distribution
(σ Unknown) 127

6.4 Confidence Interval Estimation for Categorical Variables 131

Chapter 7   Fundamentals of Hypothesis Testing **141**

7.1 The Null and Alternative Hypotheses 141

7.2 Hypothesis Testing Issues 143

7.3 Decision-Making Risks 145

7.4 Performing Hypothesis Testing 147

7.5 Types of Hypothesis Tests 148

Chapter 8   Hypothesis Testing: *Z* and *t* Tests **153**

8.1 Testing for the Difference Between Two Proportions 153

8.2 Testing for the Difference Between the Means of
Two Independent Groups 160

8.3 The Paired t Test 166

**Chapter 9   Hypothesis Testing: Chi-Square Tests and the One-Way
Analysis of Variance (ANOVA)** **179**

9.1 Chi-Square Test for Two-Way Cross-Classification Tables 179

9.2 One-Way Analysis of Variance (ANOVA): Testing for the
Differences Among the Means of More Than Two Groups 186

Chapter 10   Simple Linear Regression **207**

10.1 Basics of Regression Analysis 208

10.2 Determining the Simple Linear Regression Equation 209

10.3 Measures of Variation 217

10.4 Regression Assumptions 222

10.5 Residual Analysis 223

10.6 Inferences About the Slope 225

10.7 Common Mistakes Using Regression Analysis 228

Chapter 11   Multiple Regression **245**

11.1 The Multiple Regression Model 245

11.2 Coefficient of Multiple Determination 248

11.3 The Overall F test 249

11.4 Residual Analysis for the Multiple Regression Model 250

11.5 Inferences Concerning the Population Regression Coefficients 251

Chapter 12   Quality and Six Sigma Applications of Statistics **265**

12.1 Total Quality Management 265

12.2 Six Sigma 267

12.3 Control Charts 268

12.4 The p Chart 271

12.5 The Parable of the Red Bead Experiment: Understanding Process
Variability 276

12.6 Variables Control Charts for the Mean and Range 278

**Appendix A   Calculator and Spreadsheet Operation
and Configuration** **295**

A.C1 Calculator Operation Conventions 295

A.C2 Calculator Technical Configuration 297

A.C3 Using the A2MULREG Program 298

A.C4 Using TI Connect 298

A.S1 Spreadsheet Operation Conventions 299

A.S2 Spreadsheet Technical Configurations 299

Appendix B   Review of Arithmetic and Algebra **301**

Assessment Quiz 301

Symbols 304

Answers to Quiz 310

Appendix C   Statistical Tables **311**

Appendix D   Spreadsheet Tips **339**

CT: Chart Tips 339

FT: Function Tips 341

ATT: Analysis ToolPak Tips (Microsoft Excel only) 343

Appendix E   Advanced Techniques **347**

E.1 Using PivotTables to Create Two-Way Cross-Classification Tables 347

E.2 Using the FREQUENCY Function to Create Frequency Distributions 349

E.3 Calculating Quartiles 350

E.4 Using the LINEST Function to Calculate Regression Results 351

Appendix F   Documentation for Downloadable Files **353**

F.1 Downloadable Data Files 353

F.2 Downloadable Spreadsheet Solution Files 357

Glossary **359**

Index **367**

Acknowledgments

We would especially like to thank the staff at Financial Times/Pearson: Jim Boyd for making this book a reality, Debbie Williams for her proofreading, Paula Lowell for her copy editing, and Anne Goebel for her work in the production of this text.

We have sought to make the contents of this book as clear, accurate, and error-free as possible. We invite you to make suggestions or ask questions about the content if you think we have fallen short of our goals in any way. Please email your comments to davidlevine@davidlevinestatistics.com and include Even You Can Learn Statistics 2/e in the subject line.

About the Authors

David M. Levine is Professor Emeritus of Statistics and Computer Information Systems at Baruch College (CUNY). He received B.B.A. and M.B.A. degrees in Statistics from City College of New York and a Ph.D. degree from New York University in Industrial Engineering and Operations Research. He is nationally recognized as a leading innovator in business statistics education and is the co-author of such best-selling statistics textbooks as *Statistics for Managers Using Microsoft Excel*, *Basic Business Statistics: Concepts and Applications*, *Business Statistics: A First Course*, and *Applied Statistics for Engineers and Scientists Using Microsoft Excel and Minitab*.

He also is the author of *Statistics for Six Sigma Green Belts and Champions*, published by Financial Times–Prentice-Hall. He is coauthor of *Six Sigma for Green Belts and Champions* and *Design for Six Sigma for Green Belts and Champions* also published by Financial Times–Prentice-Hall, and *Quality Management* Third Ed., McGraw-Hill-Irwin. He is also the author of *Video Review of Statistics* and *Video Review of Probability*, both published by Video Aided Instruction. He has published articles in various journals including *Psychometrika*, *The American Statistician*, *Communications in Statistics*, *Multivariate Behavioral Research*, *Journal of Systems Management*, *Quality Progress*, and *The American Anthropologist* and has given numerous talks at American Statistical Association, Decision Sciences Institute, and Making Statistics More Effective in Schools of Business conferences. While at Baruch College, Dr. Levine received numerous awards for outstanding teaching.

David F. Stephan is an independent instructional technologist. During his more than 20 years teaching at Baruch College (CUNY), he pioneered the use of computer-equipped classrooms and interdisciplinary multimedia tools and devised techniques for teaching computer applications in a business context. The developer of PHStat2, the Pearson Education statistics add-in system for Microsoft Excel, he has collaborated with David Levine on a number of projects and is a coauthor of *Statistics for Managers Using Microsoft Excel*.

Introduction
The *Even You Can Learn Statistics* Owners Manual

In today's world, understanding statistics is more important than ever. *Even You Can Learn Statistics: A Guide for Everyone Who Has Ever Been Afraid of Statistics* can teach you the basic concepts that provide you with the knowledge to apply statistics in your life. You will also learn the most commonly used statistical methods and have the opportunity to practice those methods while using a statistical calculator or spreadsheet program.

Please read the rest of this introduction so that you can become familiar with the distinctive features of this book. You can also visit the website for this book (**www.ftpress.com/youcanlearnstatistics2e**) where you can learn more about this book as well as download files that support your learning of statistics.

Mathematics Is Always Optional!

Never mastered higher mathematics—or generally fearful of math? Not to worry, because in *Even You Can Learn Statistics* you will find that every concept is explained in plain English, without the use of higher mathematics or mathematical symbols. Interested in the mathematical foundations behind statistics? *Even You Can Learn Statistics* includes **Equation Blackboards**, stand-alone sections that present the equations behind statistical methods and complement the main material. Either way, you can learn statistics.

Learning with the Concept-Interpretation Approach

Even You Can Learn Statistics uses a **Concept-Interpretation** approach to help you learn statistics. For each important statistical concept, you will find the following:

- A **CONCEPT**, a plain language definition that uses no complicated mathematical terms
- An **INTERPRETATION**, that fully explains the concept and its importance to statistics. When necessary, these sections also discuss common

misconceptions about the concept as well as the common errors people can make when trying to apply the concept.

For simpler concepts, an **EXAMPLES** section lists real-life examples or applications of the statistical concepts. For more involved concepts, **WORKED-OUT PROBLEMS** provide a complete solution to a statistical problem—including actual spreadsheet and calculator results—that illustrate how you can apply the concept to your own situations.

Practicing Statistics While You Learn Statistics

To help you learn statistics, you should always review the worked-out problems that appear in this book. As you review them, you can practice what you have just learned by using the optional **CALCULATOR KEYS** or **SPREADSHEET SOLUTION** sections.

Calculator Keys sections provide you with the step-by-step instructions to perform statistical analysis using one of the calculators from the Texas Instruments TI-83/84 family. (You can adapt many instruction sets for use with other TI statistical calculators.)

Prefer to practice using a personal computer spreadsheet program? Spreadsheet Solution sections enable you to use Microsoft Excel or OpenOffice.org Calc 3 as you learn statistics.

If you don't want to practice your calculator or spreadsheet skills, you can examine the calculator and spreadsheet results that appear throughout the book. Many spreadsheet results are available as files that you can download for free at **www.ftpress.com/youcanlearnstatistics2e**.

Spreadsheet program users will also benefit from Appendix D, "Spreadsheet Tips" and Appendix E, "Advanced Techniques," which help teach you more about spreadsheets as you learn statistics.

And if technical issues or instructions have ever confounded your using a calculator or spreadsheet in the past, check out Appendix A, "Calculator and Spreadsheet Operation and Configuration," which details the technical configuration issues you might face and explains the conventions used in all technical instructions that appear in this book.

In-Chapter Aids

As you read a chapter, look for the following icons for extra help:

Important Point icons highlight key definitions and explanations.

File icons identify files that allow you to examine the data in selected problems. (You can download these files for free at **www.ftpress.com/youcanlearnstatistics2e.**)

Interested in the mathematical foundations of statistics? Then look for the Interested in Math? icons throughout the book. But remember, you can skip any or all of the math sections without losing any comprehension of the statistical methods presented, because math is always optional in this book!

End-of-Chapter Features

At the end of most chapters of *Even You Can Learn Statistics* you can find the following features, which you can review to reinforce your learning.

Important Equations

The **Important Equations** sections present all of the important equations discussed in the chapter. Even if you are not interested in the mathematics of the statistical methods and have skipped the Equation Blackboards in the book, you can use these lists for reference and later study.

One-Minute Summaries

One-Minute Summaries are a quick review of the significant topics of a chapter in outline form. When appropriate, the summaries also help guide you to make the right decisions about applying statistics to the data you seek to analyze.

Test Yourself

The **Test Yourself** sections offer a set of short-answer questions and problems that enable you to review and test yourself (with answers provided) to see how much you have retained of the concepts presented in a chapter.

New to the Second Edition

The following features are new to this second edition:

- Problems (and answers) are included as part of the Test Yourself sections at the end of chapters.
- The book has expanded coverage of the use of spreadsheet programs for solving statistical programs.
- A new chapter (Chapter 11, "Multiple Regression") covers the essentials of multiple regression that expands on the concepts of simple linear regression covered in Chapter 10, "Simple Linear Regression."
- Many new and revised examples are included throughout the book.

Summary

Even You Can Learn Statistics can help you whether you are studying statistics as part of a formal course or just brushing up on your knowledge of statistics for a specific analysis. Be sure to visit the website for this book (**www.ftpress.com/youcanlearnstatistics2e**) and feel free to contact the authors via email at davidlevine@davidlevinestatistics.com; include Even You Can Learn Statistics 2/e in the subject line if you have any questions about this book.

Fundamentals of Statistics

1.1 The First Three Words of Statistics
1.2 The Fourth and Fifth Words
1.3 The Branches of Statistics
1.4 Sources of Data
1.5 Sampling Concepts
1.6 Sample Selection Methods
One-Minute Summary
Test Yourself

Every day, the media uses numbers to describe or analyze our world:

- "Americans Gulping More Bottled Water"—The annual per capita consumption of bottled water has increased from 18.8 gallons in 2001 to 28.3 gallons in 2006.

- "Summer Sports Are Among the Safest"—Researchers at the Centers for Disease Control and Prevention report that the most dangerous outdoor activity is snowboarding. The injury rate for snowboarding is higher than for all the summer pastimes combined.

- "Reducing Prices Has a Different Result at Barnes & Noble than at Amazon"—A study reveals that raising book prices by 1% reduced sales by 4% at BN.com, but reduced sales by only 0.5% at Amazon.com.

- "Four out of five dentists recommend…"—A typically encountered advertising claim for chewing gum or oral hygiene products.

You can make better sense of the numbers you encounter if you learn to understand statistics. **Statistics**, a branch of mathematics, uses procedures that allow you to correctly analyze the numbers. These procedures, or **statistical methods**, transform numbers into useful information that you can use when making decisions about the numbers. Statistical methods can also tell you the known risks associated with making a decision as well as help you make more consistent judgments about the numbers.

Learning statistics requires you to reflect on the significance and the importance of the results to the decision-making process you face. This statistical interpretation means knowing when to ignore results because they are misleading, are produced by incorrect methods, or just restate the obvious, as in "100% of the authors of this book are named 'David.'"

In this chapter, you begin by learning five basic words—*population*, *sample*, *variable*, *parameter*, and *statistic* (singular)—that identify the fundamental concepts of statistics. These five words, and the other concepts introduced in this chapter, help you explore and explain the statistical methods discussed in later chapters.

1.1   The First Three Words of Statistics

You've already learned that statistics is about analyzing things. Although *numbers* was the word used to represent things in the opening of this chapter, the first three words of statistics, *population*, *sample*, and *variable*, help you to better identify what you analyze with statistics.

Population

CONCEPT   All the members of a group about which you want to draw a conclusion.

EXAMPLES   All U.S. citizens who are currently registered to vote, all patients treated at a particular hospital last year, the entire daily output of a cereal factory's production line.

Sample

CONCEPT   The part of the population selected for analysis.

EXAMPLES   The registered voters selected to participate in a recent survey concerning their intention to vote in the next election, the patients selected to fill out a patient satisfaction questionnaire, 100 boxes of cereal selected from a factory's production line.

Variable

CONCEPT   A characteristic of an item or an individual that will be analyzed using statistics.

EXAMPLES   Gender, the party affiliation of a registered voter, the household income of the citizens who live in a specific geographical area, the publishing category (hardcover, trade paperback, mass-market paperback, textbook) of a book, the number of televisions in a household.

INTERPRETATION   All the variables taken together form the data of an analysis. Although people often say that they are analyzing their data, they are, more precisely, analyzing their variables. (Consistent to everyday usage, the authors use these terms interchangeably throughout this book.)

You should distinguish between a variable, such as gender, and its value for an individual, such as male. An **observation** is all the values for an individual item in the sample. For example, a survey might contain two variables, gender and age. The first observation might be male, 40. The second observation might be female, 45. The third observation might be female, 55. A variable is sometimes known as a column of data because of the convention of entering each observation as a unique row in a table of data. (Likewise, some people refer to an observation as a row of data.)

Variables can be divided into the following types:

	Categorical Variables	**Numerical Variables**
Concept	The values of these variables are selected from an established list of categories.	The values of these variables involve a counted or measured value.
Subtypes	None	**Discrete** values are counts of things.
		Continuous values are measures and any value can theoretically occur, limited only by the precision of the measuring process.
Examples	Gender, a variable that has the categories "male" and "female."	The number of people living in a household, a discrete numerical variable.
	Academic major, a variable that might have the categories "English," "Math," "Science," and "History," among others.	The time it takes for someone to commute to work, a continuous variable.

All variables should have an operational definition—that is, a universally accepted meaning that is understood by all associated with an analysis. Without operational definitions, confusion can occur. A famous example of such confusion was the tallying of votes in Florida during the 2000 U.S. presidential election in which, at various times, nine different definitions of a valid ballot were used. (A later analysis[1] determined that three of these definitions, including one pursued by Al Gore, led to margins of victory for George Bush that ranged from 225 to 493 votes and that the six others, including one pursued by George Bush, led to margins of victory for Al Gore that ranged from 42 to 171 votes.)

1.2 The Fourth and Fifth Words

After you know what you are analyzing, or, using the words of Section 1.1, after you have identified the variables from the population or sample under study, you can define the **parameters** and **statistics** that your analysis will determine.

Parameter

CONCEPT   A numerical measure that describes a variable (characteristic) of a population.

EXAMPLES   The percentage of all registered voters who intend to vote in the next election, the percentage of all patients who are very satisfied with the care they received, the mean weight of all the cereal boxes produced at a factory on a particular day.

Statistic

CONCEPT   A numerical measure that describes a variable (characteristic) of a sample (part of a population).

EXAMPLES   The percentage of registered voters in a sample who intend to vote in the next election, the percentage of patients in a sample who are very satisfied with the care they received, the mean weight of a sample of cereal boxes produced at a factory on a particular day.

INTERPRETATION   Calculating statistics for a sample is the most common activity because collecting population data is impractical in most actual decision-making situations.

---

[1] J. Calmes and E. P. Foldessy, "In Election Review, Bush Wins with No Supreme Court Help," *Wall Street Journal*, November 12, 2001, A1, A14.

1.3 The Branches of Statistics

You can use parameters and statistics either to describe your variables or to reach conclusions about your data. These two uses define the two branches of statistics: **descriptive statistics** and **inferential statistics**.

Descriptive Statistics

CONCEPT   The branch of statistics that focuses on collecting, summarizing, and presenting a set of data.

EXAMPLES   The mean age of citizens who live in a certain geographical area, the mean length of all books about statistics, the variation in the weight of 100 boxes of cereal selected from a factory's production line.

INTERPRETATION   You are most likely to be familiar with this branch of statistics because many examples arise in everyday life. Descriptive statistics serves as the basis for analysis and discussion in fields as diverse as securities trading, the social sciences, government, the health sciences, and professional sports. Descriptive methods can seem deceptively easy to apply because they are often easily accessible in calculating and computing devices. However, this easiness does not mean that descriptive methods are without their pitfalls, as Chapter 2, "Presenting Data in Charts and Tables," and Chapter 3, "Descriptive Statistics," explain.

Inferential Statistics

CONCEPT   The branch of statistics that analyzes sample data to reach conclusions about a population.

EXAMPLE   A survey that sampled 1,264 women found that 45% of those polled considered friends or family as their most trusted shopping advisers and only 7% considered advertising as their most trusted shopping adviser. By using methods discussed in Section 6.4, you can use these statistics to draw conclusions about the population of all women.

INTERPRETATION   When you use inferential statistics, you start with a hypothesis and look to see whether the data are consistent with that hypothesis. This deeper level of analysis means that inferential statistical methods can be easily misapplied or misconstrued, and that many inferential methods require a calculating or computing device. (Chapters 6 through 9 discuss some of the inferential methods that you will most commonly encounter.)

1.4 Sources of Data

You begin every statistical analysis by identifying the source of the data. Among the important sources of data are **published sources**, **experiments**, and **surveys**.

Published Sources

CONCEPT   Data available in print or in electronic form, including data found on Internet websites. Primary data sources are those published by the individual or group that collected the data. Secondary data sources are those compiled from primary sources.

EXAMPLE   Many U.S. federal agencies, including the Census Bureau, publish primary data sources that are available at the **www.fedstats.gov** website. Business news sections of daily newspapers commonly publish secondary source data compiled by business organizations and government agencies.

INTERPRETATION   You should always consider the possible bias of the publisher and whether the data contain all the necessary and relevant variables when using published sources. Remember, too, that anyone can publish data on the Internet.

Experiments

CONCEPT   A study that examines the effect on a variable of varying the value(s) of another variable or variables, while keeping all other things equal. A typical experiment contains both a treatment group and a control group. The treatment group consists of those individuals or things that receive the treatment(s) being studied. The control group consists of those individuals or things that do not receive the treatment(s) being studied.

EXAMPLE   Pharmaceutical companies use experiments to determine whether a new drug is effective. A group of patients who have many similar characteristics is divided into two subgroups. Members of one group, the treatment group, receive the new drug. Members of the other group, the control group, often receive a placebo, a substance that has no medical effect. After a time period, statistics about each group are compared.

INTERPRETATION   Proper experiments are either single-blind or double-blind. A study is a single-blind experiment if only the researcher conducting the study knows the identities of the members of the treatment and control groups. If neither the researcher nor study participants know who is in the treatment group and who is in the control group, the study is a double-blind experiment.

When conducting experiments that involve placebos, researchers also have to consider the placebo effect—that is, whether people in the control group will improve because they believe they are getting a real substance that is intended to produce a positive result. When a control group shows as much improvement as the treatment group, a researcher can conclude that the placebo effect is a significant factor in the improvements of both groups.

Surveys

CONCEPT   A process that uses questionnaires or similar means to gather values for the responses from a set of participants.

EXAMPLES   The decennial U.S. census mail-in form, a poll of likely voters, a website instant poll or "question of the day."

INTERPRETATION   Surveys are either **informal**, open to anyone who wants to participate; **targeted**, directed toward a specific group of individuals; or include people chosen at random. The type of survey affects how the data collected can be used and interpreted.

1.5 Sampling Concepts

In the definition of **statistic** in Section 1.2, you learned that calculating statistics for a sample is the most common activity because collecting population data is usually impractical. Because samples are so commonly used, you need to learn the concepts that help identify all the members of a population and that describe how samples are formed.

Frame

CONCEPT   The list of all items in the population from which the sample will be selected.

EXAMPLES   Voter registration lists, municipal real estate records, customer or human resource databases, directories.

INTERPRETATION   Frames influence the results of an analysis, and using different frames can lead to different conclusions. You should always be careful to make sure your frame completely represents a population; otherwise, any sample selected will be biased, and the results generated by analyses of that sample will be inaccurate.

Sampling

CONCEPT   The process by which members of a *population* are selected for a *sample*.

EXAMPLES   Choosing every fifth voter who leaves a polling place to interview, selecting playing cards randomly from a deck, polling every tenth visitor who views a certain website today.

INTERPRETATION   Some sampling techniques, such as an "instant poll" found on a web page, are naturally suspect as such techniques do not depend on a well-defined frame. The sampling technique that uses a well-defined frame is **probability sampling**.

Probability Sampling

CONCEPT   A sampling process that considers the chance of selection of each item. Probability sampling increases your chance that the sample will be representative of the population.

EXAMPLES   The registered voters selected to participate in a recent survey concerning their intention to vote in the next election, the patients selected to fill out a patient-satisfaction questionnaire, 100 boxes of cereal selected from a factory's production line.

INTERPRETATION   You should use probability sampling whenever possible, because *only* this type of sampling enables you to apply inferential statistical methods to the data you collect. In contrast, you should use nonprobability sampling, in which the chance of occurrence of each item being selected is not known, to obtain rough approximations of results at low cost or for small-scale, initial, or pilot studies that will later be followed up by a more rigorous analysis. Surveys and polls that invite the public to call in or answer questions on a web page are examples of nonprobability sampling.

Simple Random Sampling

CONCEPT   The probability sampling process in which every individual or item from a population has the same chance of selection as every other individual or item. Every possible sample of a certain size has the same chance of being selected as every other sample of that size.

EXAMPLES   Selecting a playing card from a shuffled deck or using a statistical device such as a table of random numbers.

INTERPRETATION   Simple random sampling forms the basis for other random sampling techniques. The word *random* in this phrase requires clarification. In this phrase, *random* means no repeating patterns—that is, in a given sequence, a given pattern is equally likely (or unlikely). It does not refer to the most commonly used meaning of "unexpected" or "unanticipated" (as in "random acts of kindness").

Other Probability Sampling Methods

Other, more complex, sampling methods are also used in survey sampling. In a stratified sample, the items in the frame are first subdivided into separate subpopulations, or strata, and a simple random sample is selected within each of the strata. In a cluster sample, the items in the frame are divided into several clusters so that each cluster is representative of the entire population. A random sampling of clusters is then taken, and all the items in each selected cluster or a sample from each cluster are then studied.

1.6 Sample Selection Methods

Proper sampling can be done either with or without replacement of the items being selected.

Sampling with Replacement

CONCEPT   A sampling method in which each selected item is returned to the frame from which it was selected so that it has the same probability of being selected again.

EXAMPLE   Selecting items from a fishbowl and returning each item to it after the selection is made.

Sampling Without Replacement

CONCEPT   A sampling method in which each selected item is not returned to the frame from which it was selected. Using this technique, an item can be selected no more than one time.

EXAMPLES   Selecting numbers in state lottery games, selecting cards from a deck of cards during games of chance such as blackjack or poker.

INTERPRETATION   Sampling without replacement means that an item can be selected no more than one time. You should choose sampling without replacement instead of sampling with replacement because statisticians generally consider the former to produce more desirable samples.

calculator keys

Entering Data

You enter the data values of a variable into one of six predefined **list variables**: L1 through L6. Your method of data entry varies, depending on the number of values to enter and personal preferences.

For small sets of values, you enter the values separated by commas as follows:

- Press [**2nd**][(] and then type the values separated by commas. If your list is longer than the width of the screen, the list wraps to the next line like so:

```
{39,29,43,52,39,
44,40,31,44,35█
```

- Press [**2nd**][)][**STO▶**] and then type the variable name and press [**ENTER**].

To store the values in variable L1, press [**2nd**][)][**STO▶**] then [**2nd**][1][**Enter**]. ([**2nd**][1] types L1, [**2nd**][2] types L2, and so forth.) Your calculator displays the variable name and one line's worth of values, separated by spaces, followed by an ellipsis if the entire list of values cannot be shown on one line.

```
{39,29,43,52,39,
44,40,31,44,35}→
L1
{39 29 43 52 39…
█
```

For larger sets of data values, consider using an editor. For a calculator not connected to a computer, use the calculator's statistical list editor:

- Press [**STAT**].

- Select **1:Edit** and press [**ENTER**].

- In the editor's six-column table (one column for each list variable), use the cursor keys to move through the table and make entries. End every entry by pressing [ENTER].

- When you are finished, press [2nd][MODE] to quit the editor.

While you are in the editor, you can move back in the column and make changes to a previously entered value. If you need to erase all the values of a column (to reuse a list variable), move the cursor to the name of the list variable (at the top of its column) and press [CLEAR][ENTER].

If your calculator is connected to a computer, you can use the TI DataEditor component of the TI Connect program (see Section A.C4). To enter a list using the DataEditor, open TI Connect, click the **TI DataEditor** icon, and in the DataEditor window:

- Select **File → New → List**.

- Select **File → Properties** and specify your calculator type and the (list) variable name in the Variable Properties dialog box.

- Enter the data values in the spreadsheet-like column.

- When you are finished, click the **Send File** icon to transfer the variable data to your calculator.

The following illustrations show the calculator's statistical list editor and the DataEditor window, respectively, after all the values of the earlier example have been entered.

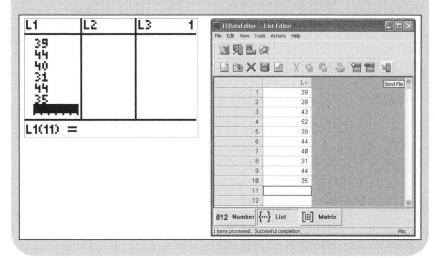

spreadsheet solution

Entering Data

Enter the data values of a variable in a blank column of a worksheet. Use the row 1 cell for the variable name.

To create a new file (workbook) for your entries, do the following:

- In Microsoft Excel versions 2007 or later, click the **Office Button**, select **New**, and in the New Workbook dialog box, double-click the **Blank Workbook** icon.

- In earlier Excel versions, select **File → New** and select **Blank workbook** from a New Workbook task pane, or select the **Workbook** icon if the **New** dialog box appears.

- In OpenOffice.org Calc 3, select **File → New → Spreadsheet**.

To save your work, select **Office Button → Save As** in Excel 2007 or later. Select **File → Save As** in earlier Excel versions and OpenOffice.org Calc 3.

important point

In this book, consecutive menu selections in spreadsheet programs are shown linked with this symbol: →. When you read a phrase such as **File → Save As**, you should interpret the phrase as "select **File** from the menu list near the top of the spreadsheet window and then select **Save As** from the dropdown menu that appears."

One-Minute Summary

To understand statistics, you must first master the basic vocabulary presented in this chapter. You have also been introduced to data collection, the various sources of data, sampling methods, as well as the types of variables used in statistical analysis. The remaining chapters of this book focus on four important reasons for learning statistics:

- To present and describe information (Chapters 2 and 3)
- To reach conclusions about populations based only on sample results (Chapters 4 through 9)
- To develop reliable forecasts (Chapters 10 and 11)
- To improve processes (Chapter 12)

Test Yourself

1. The portion of the population that is selected for analysis is called:
   (a) a sample

   (b) a frame

   (c) a parameter

   (d) a statistic

2. A summary measure that is computed from only a sample of the population is called:
   (a) a parameter

   (b) a population

   (c) a discrete variable

   (d) a statistic

3. The height of an individual is an example of a:
   (a) discrete variable

   (b) continuous variable

   (c) categorical variable

   (d) constant

4. The body style of an automobile (sedan, coupe, wagon, and so on) is an example of a:
   (a) discrete variable

   (b) continuous variable

   (c) categorical variable

   (d) constant

5. The number of credit cards in a person's wallet is an example of a:
   (a) discrete variable

   (b) continuous variable

   (c) categorical variable

   (d) constant

6. Statistical inference occurs when you:
   (a) compute descriptive statistics from a sample

   (b) take a complete census of a population

   (c) present a graph of data

   (d) take the results of a sample and reach conclusions about a population

7. The human resources director of a large corporation wants to develop a dental benefits package and decides to select 100 employees from a list of all 5,000 workers in order to study their preferences for the various

components of a potential package. All the employees in the corporation constitute the _____.
   (a) sample
   (b) population
   (c) statistic
   (d) parameter

8. The human resources director of a large corporation wants to develop a dental benefits package and decides to select 100 employees from a list of all 5,000 workers in order to study their preferences for the various components of a potential package. The 100 employees who will participate in this study constitute the _____.
   (a) sample
   (b) population
   (c) statistic
   (d) parameter

9. Those methods that involve collecting, presenting, and computing characteristics of a set of data in order to properly describe the various features of the data are called:
   (a) statistical inference
   (b) the scientific method
   (c) sampling
   (d) descriptive statistics

10. Based on the results of a poll of 500 registered voters, the conclusion that the Democratic candidate for U.S. president will win the upcoming election is an example of:
    (a) inferential statistics
    (b) descriptive statistics
    (c) a parameter
    (d) a statistic

11. A numerical measure that is computed to describe a characteristic of an entire population is called:
    (a) a parameter
    (b) a population
    (c) a discrete variable
    (d) a statistic

12. You were working on a project to examine the value of the American dollar as compared to the English pound. You accessed an Internet site where you obtained this information for the past 50 years. Which method of data collection were you using?

(a) published sources

(b) experimentation

(c) surveying

13. Which of the following is a discrete variable?
    (a) The favorite flavor of ice cream of students at your local elementary school
    (b) The time it takes for a certain student to walk to your local elementary school
    (c) The distance between the home of a certain student and the local elementary school
    (d) The number of teachers employed at your local elementary school

14. Which of the following is a continuous variable?
    (a) The eye color of children eating at a fast-food chain
    (b) The number of employees of a branch of a fast-food chain
    (c) The temperature at which a hamburger is cooked at a branch of a fast-food chain
    (d) The number of hamburgers sold in a day at a branch of a fast-food chain

15. The number of cars that arrive per hour at a parking lot is an example of:
    (a) a categorical variable
    (b) a discrete variable
    (c) a continuous variable
    (d) a statistic

Answer True or False:

16. The possible responses to the question, "How long have you been living at your current residence?" are values from a continuous variable.

17. The possible responses to the question, "How many times in the past three months have you visited a museum?" are values from a discrete variable.

Fill in the blank:

18. An insurance company evaluates many variables about a person before deciding on an appropriate rate for automobile insurance. The number of accidents a person has had in the past three years is an example of a _____ variable.

19. An insurance company evaluates many variables about a person before deciding on an appropriate rate for automobile insurance. The distance a person drives in a day is an example of a _____ variable.

20. An insurance company evaluates many variables about a person before deciding on an appropriate rate for automobile insurance. A person's marital status is an example of a _____ variable.

21. A numerical measure that is computed from only a sample of the population is called a _____.

22. The portion of the population that is selected for analysis is called the _____.

23. A college admission application includes many variables. The number of advanced placement courses the student has taken is an example of a _____ variable.

24. A college admission application includes many variables. The gender of the student is an example of a _____ variable.

25. A college admission application includes many variables. The distance from the student's home to the college is an example of a _____ variable.

Answers to Test Yourself

1. a
2. d
3. b
4. c
5. a
6. d
7. b
8. a
9. d
10. a
11. a
12. a
13. d

14. c
15. b
16. True
17. True
18. discrete
19. continuous
20. categorical
21. statistic
22. sample
23. discrete
24. categorical
25. continuous

References

1. Berenson, M. L., D. M. Levine, and T. C. Krehbiel. *Basic Business Statistics: Concepts and Applications*, Eleventh Edition. Upper Saddle River, NJ: Prentice Hall, 2009.

2. Cochran, W. G. *Sampling Techniques*, Third Edition. New York: John Wiley & Sons, 1977.

3. D. M. Levine. *Statistics for Six Sigma Green Belts with Minitab and JMP*. Upper Saddle River, NJ: Financial Times – Prentice Hall, 2006.

4. Levine, D. M., T. C. Krehbiel, and M. L. Berenson. *Business Statistics: A First Course*, Fifth Edition. Upper Saddle River, NJ: Prentice Hall, 2010.

5. Levine, D. M., D. Stephan, T. C. Krehbiel, and M. L. Berenson. *Statistics for Managers Using Microsoft Excel*, Fifth Edition. Upper Saddle River, NJ: Prentice Hall, 2008.

6. Levine, D. M., P. P. Ramsey, and R. K. Smidt, *Applied Statistics for Engineers and Scientists Using Microsoft Excel and Minitab*. Upper Saddle River, NJ: Prentice Hall, 2001.

2

Presenting Data in Charts and Tables

2.1 Presenting Categorical Variables
2.2 Presenting Numerical Variables
2.3 Misusing Charts
One-Minute Summary
Test Yourself

In an information-overloaded world, you need to present information effectively. You can present categorical and numerical data efficiently using charts and tables. Reading this chapter can help you learn to select and develop charts and tables for each type of data.

2.1 Presenting Categorical Variables

You present a categorical variable by first sorting variable values according to the categories of the variable. Then you place the count, amount, or percentage (part of the whole) of each category into a summary table or into one of several types of charts.

The Summary Table

CONCEPT A two-column table in which category names are listed in the first column and the count, amount, or percentage of values are listed in a second column. Sometimes, additional columns present the same data in more than one way (for example, as counts and percentages).

EXAMPLE   The results of a survey that asked adults how they pay their monthly bills can be presented using a summary table:

Form of Payment	Percentage (%)
Cash	15
Check	54
Electronic/online	28
Other/don't know	3

Source: Data extracted from *USA Today Snapshots*, "How Adults Pay Monthly Bills," *USA Today*, October 4, 2007, p. 1.

INTERPRETATION   Summary tables enable you to see the big picture about a set of data. In this example, you can conclude that more than half the people pay by check and almost 75% either pay by check or by electronic/online forms of payment.

The Bar Chart

CONCEPT   A chart containing rectangles ("bars") in which the length of each bar represents the count, amount, or percentage of responses of one category.

EXAMPLE   This percentage bar chart presents the data of the summary table discussed in the previous example:

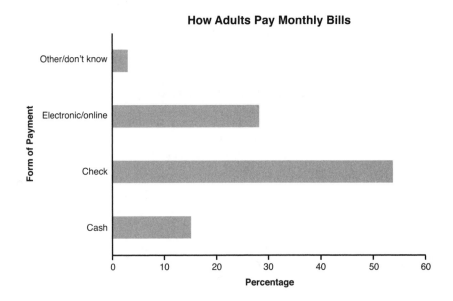

INTERPRETATION   A bar chart is better than a summary table at making the point that the category "pay by check" is the single largest category for this example. For most people, scanning a bar chart is easier than scanning a column of numbers in which the numbers are unordered, as they are in the bill payment summary table.

The Pie Chart

CONCEPT   A circle chart in which wedge-shaped areas—pie slices—represent the count, amount, or percentage of each category and the entire circle ("pie") represents the total.

EXAMPLE   This pie chart presents the data of the summary table discussed in the preceding two examples:

How Adults Pay Monthly Bills

Other/don't know
3%

Cash
15%

Electronic/online
28%

Check
54%

INTERPRETATION   The pie chart enables you to see each category's portion of the whole. You can see that most of the adults pay their monthly bills by check or electronic/online, a small percentage pay with cash, and that hardly anyone paid using another form of payment or did not know how they paid.

Although you can probably create most of your pie charts using electronic means, you can also create a pie chart using a protractor to divide up a drawn circle. To create a pie chart in this way, first calculate percentages for each category. Then multiply each percentage by 360, the number of degrees in a circle, to get the number of degrees for the arc (part of circle) that represents each category's pie slice. For example, for the "pay by check" category, multiply 54% by 360 degrees to get 194.4 degrees. Mark the endpoints of this arc on the circle using the protractor, and draw lines from the endpoints to the center of the circle. (If you draw your circle using a compass the center of the circle can be easily identified.)

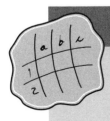

spreadsheet solution

Bar and Pie Charts

Chapter 2 Bar and **Chapter 2 Pie** contain examples of a bar and pie chart, respectively. Experiment with each chart by entering your own values in column B.

Spreadsheet Tips CT1 and CT2 (see Appendix D) explain how to further modify these charts.

If you are a knowledgeable spreadsheet user, you can create your own charts from scratch. Spreadsheet Tip CT3 (see Appendix D) discusses the general steps for creating charts.

important point

The Pareto Chart

CONCEPT   A special type of bar chart that presents the counts, amounts, or percentages of each category in descending order left to right, and also contains a superimposed plotted line that represents a running cumulative percentage.

EXAMPLE

Computer Keyboards Defects for a Three-Month Period.

Defect	Frequency	Percentage
Black spot	413	6.53%
Damage	1,039	16.43%
Jetting	258	4.08%
Pin mark	834	13.19%
Scratches	442	6.99%
Shot mold	275	4.35%
Silver streak	413	6.53%
Sink mark	371	5.87%

Defect	Frequency	Percentage
Spray mark	292	4.62%
Warpage	1,987	31.42%
Total*	6,324	100.01%

*Total percentage equals 100.01 due to rounding.

Source: Data extracted from U. H. Acharya and C. Mahesh, "Winning Back the Customer's Confidence: A Case Study on the Application of Design of Experiments to an Injection-Molding Process," *Quality Engineering*, 11, 1999, 357–363.

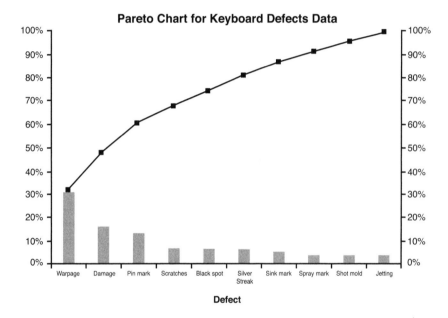

This Pareto chart uses the data of the table that immediately precedes it to highlight the causes of computer keyboard defects manufactured during a three-month period.

INTERPRETATION When you have many categories, a Pareto chart enables you to focus on the most important categories by visually separating the "vital few" from the "trivial many" categories. For the keyboard defects data, the Pareto chart shows that two categories, warpage and damage, account for nearly one-half of all defects, and that those two categories combined with the pin mark category account for more than 60% of all defects.

spreadsheet solution

Pareto Charts

Chapter 2 Pareto contains an example of a Pareto chart. Experiment with this chart by typing your own set of values—in descending order—in column B, rows 2 through 11. (Do not alter the entries in row 12 or columns C and D.)

Spreadsheet Tip CT4 (see Appendix D) summarizes how to create a Pareto chart from scratch.

Two-Way Cross-Classification Table

CONCEPT   A multicolumn table that presents the count or percentage of responses for two categorical variables. In a two-way table, the categories of one of the variables form the rows of the table, while the categories of the second variable form the columns. The "outside" of the table contains a special row and a special column that contain the totals. Cross-classification tables are also known as cross-tabulation tables.

EXAMPLES

Counts of Particles Found Cross-Classified by Wafer Condition

		Wafer Condition		
		Good	Bad	Total
Particles	Yes	14	36	50
Found	No	320	80	400
	Total	334	116	450

This two-way cross-classification table summarizes the results of a manufacturing plant study that investigated whether particles found on silicon wafers affected the condition of a wafer. Tables showing row percentages, column percentages, and overall total percentages follow.

Row Percentages Table

		Wafer Condition		
		Good	Bad	Total
Particles	Yes	28.0	72.0	100.0
Found	No	80.0	20.0	100.0
	Total	74.2	25.8	100.0

Column Percentages Table

		Wafer Condition			
		Good	Bad		Total
Particles	Yes	4.2	31.0		11.1
Found	No	95.8	69.0		88.9
	Total	100.0	100.0		100.0

Overall Total Percentages Table

		Wafer Condition			
		Good	Bad		Total
Particles	Yes	3.1	8.0		11.1
Found	No	71.1	17.8		88.9
	Total	74.2	25.8		100.0

INTERPRETATION   The simplest two-way table has two rows and two columns in its inner part. Each inner cell represents the count or percentage of a pairing, or cross-classifying, of categories from each variable. Sometimes additional rows and columns present the percentages of the overall total, the percentages of the row total, and the percentages of the column total for each row and column combination.

		Column Variable		Total
		1	**2**	**Total**
	1	Count or percentage for row 1, column 1	Count or percentage for row 1, column 2	Total for row 1
Row Variable	**2**	Count or percentage for row 2, column 1	Count or percentage for row 2, column 2	Total for row 2
Total		Total for column 1	Total for column 2	Overall total

Two-way tables can reveal the combination of values that occur most often in data. In this example, the tables reveal that bad wafers are much more likely to have particles than the good wafers. Because the number of good and bad wafers was unequal in this example, you can see this pattern best in the Row Percentages table. That table shows that nearly three-quarters of the wafers that had particles were bad, but only 20% of wafers that did not have particles were bad.

PivotTables (DataPilot Tables in OpenOffice.org Calc) create worksheet summary tables from sample data and are a good way of creating a two-way table from sample data. Section E.1 in Appendix E, "Advanced Techniques" discusses how to create these tables.

spreadsheet solution

Two-Way Tables

Chapter 2 Two-Way contains the counts of particles found cross-classified by wafer condition as a simple two-way table.

Chapter 2 Two-Way PivotTable presents the same table as a PivotTable that summarizes a sample of 450 wafers.

2.2 Presenting Numerical Variables

You present numerical variables either in tables or charts. To create a table, you first establish groups that represent separate ranges of values and then place each value into the appropriate group. To create a chart, you use the groups from the table.

Many times you want to do both, and this section reviews some commonly used frequency and percentage distribution tables and the histogram chart. The frequency and percentage distributions and the histogram are among the many tables and charts that enable you to accomplish these tasks.

The Frequency and Percentage Distribution

CONCEPT   A table of grouped numerical data that contains the names of each group in the first column, the counts (frequencies) of each group in the second column, and the percentages of each group in the third column. This table can also appear as a two-column table that shows either the frequencies or the percentages.

EXAMPLE   The following Fan Cost Index data shown provides the cost in a recent year of attending an NBA professional basketball league game, including four tickets, two beers, four soft drinks, four hot dogs, two game programs, two caps, and the parking fee for one car for each of the 29 teams.

NBACost

Team	Fan Cost Index	Team	Fan Cost Index
Atlanta	244.48	Minnesota	231.38
Boston	358.72	New Jersey	328.90
Charlotte	196.90	New Orleans	182.30
Chicago	335.00	New York	394.52
Cleveland	317.90	Orlando	229.82
Dallas	339.23	Philadelphia	269.48
Denver	271.16	Phoenix	302.04
Detroit	282.00	Portland	251.86
Golden State	206.52	Sacramento	318.30
Houston	270.94	San Antonio	303.79
Indiana	250.57	Seattle	229.50
LA Clippers	317.00	Toronto	320.47
LA Lakers	453.95	Utah	235.75
Memphis	228.28	Washington	194.56
Miami	339.20		

Source: Data extracted from TeamMarketing.com, November 2007.

Remember that the file icon identifies a file that you can download for free from the website for this book (**www.ftpress.com/youcanlearnstatistics2e**). See Appendix F for more information about downloading files.

The frequency and percentage distribution for the NBA Fan Cost Index is as follows:

Fan Cost Index ($)	Frequency	Percentage
150 to under 200	3	10.34%
200 to under 250	7	24.14%
250 to under 300	6	20.69%
300 to under 350	10	34.48%
350 to under 400	2	6.90%
400 to under 450	0	0.00%
450 to under 500	1	3.45%
	29	100.00%

INTERPRETATION   Frequency and percentage distributions enable you to quickly determine differences among the many groups of values. In this example, you can quickly see that most of the fan cost indexes are between $200 and $350, and that very few fan cost indexes are either below $200 or above $350.

You need to be careful in forming groups for distributions because the ranges of the group affect how you perceive the data. For example, had the fan cost indexes been grouped into only two groups, below $300 and $300 and above, you would not be able to see any pattern in the data.

Histogram

CONCEPT   A special bar chart for grouped numerical data in which the frequencies or percentages in each group of numerical data are represented as individual bars on the vertical *Y* axis and the variable is plotted on the horizontal *X* axis. In a histogram, in contrast to a bar chart of categorical data, no gaps exist between adjacent bars.

EXAMPLE   The following histogram presents the fan cost index data of the preceding example. The values below the bars (175, 225, 275, 325, 375, 425, 475) are **midpoints**, the approximate middle value for each group of data. As with the frequency and percentage distributions, you can quickly see that very few fan cost indexes are either below $200 or above $350.

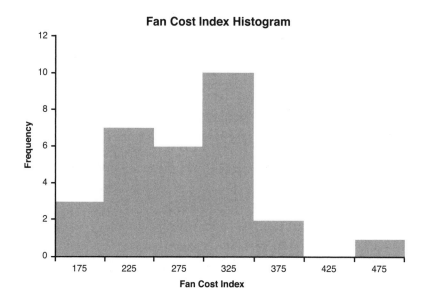

INTERPRETATION   Histograms reveal the overall shape of the frequencies in the groups. Histograms are considered symmetric if each side of the chart is an approximate mirror image of the other side. (The histogram of this example has more values in the lower portion than in the upper portion so it is considered to be skewed or non-symmetric.)

spreadsheet solution

Frequency Distributions and Histograms

Chapter 2 Histogram contains a frequency distribution and histogram for the Fan Cost Index (NBACost) data. You can experiment by typing your own set of values in column B, rows 3 through 11. (Do not alter the entries in other cells.)

Spreadsheet Tips CT5 and CT6 (in Appendix D) discuss how you can create frequency distributions and histograms from scratch.

The Time-Series Plot

CONCEPT   A chart in which each point represents the value of a numerical variable at a specific time. By convention, the X axis (the horizontal axis) always represents units of time, and the Y axis (the vertical axis) always represents units of the variable.

EXAMPLE   The following data shows the mean hotel room rate in dollars for 1996 through 2006:

Hotels

Year	Rate($)	Year	Rate($)
1996	70.63	2001	88.27
1997	75.31	2002	83.54
1998	78.62	2003	82.52
1999	81.33	2004	86.23
2000	85.89	2005	90.88
		2006	97.78

Source: Data extracted from *USA Today Snapshots*, February 13, 2008, p. 1A.

The time-series plot of these data follows.

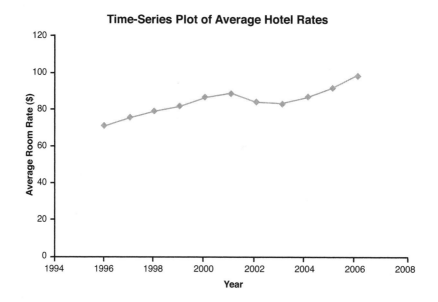

Time-Series Plot of Average Hotel Rates

INTERPRETATION   Time-series plots can reveal patterns over time, patterns that you might not see when looking at a long list of numerical values. In this example, the plot reveals that mean hotel rates were generally rising between 1996 and 2006, but declined in the years immediately after 2001.

The Scatter Plot

CONCEPT   A chart that plots the values of two numerical variables for each observation. In a scatter plot, the X axis (the horizontal axis) always represents units of one variable, and the Y axis (the vertical axis) always represents units of the second variable.

EXAMPLE   The following data tabulates the labor hours used and the cubic feet of material moved for 36 moving jobs:

Moving

Job	Labor hours	Cubic feet	Job	Labor hours	Cubic feet
M-1	24.00	545	M-19	25.00	557
M-2	13.50	400	M-20	45.00	1,028
M-3	26.25	562	M-21	29.00	793
M-4	25.00	540	M-22	21.00	523
M-5	9.00	220	M-23	22.00	564
M-6	20.00	344	M-24	16.50	312
M-7	22.00	569	M-25	37.00	757
M-8	11.25	340	M-26	32.00	600

Job	Labor hours	Cubic feet	Job	Labor hours	Cubic feet
M-9	50.00	900	M-27	34.00	796
M-10	12.00	285	M-28	25.00	577
M-11	38.75	865	M-29	31.00	500
M-12	40.00	831	M-30	24.00	695
M-13	19.50	344	M-31	40.00	1,054
M-14	18.00	360	M-32	27.00	486
M-15	28.00	750	M-33	18.00	442
M-16	27.00	650	M-34	62.50	1,249
M-17	21.00	415	M-35	53.75	995
M-18	15.00	275	M-36	79.50	1,397

The scatter plot of these data follows.

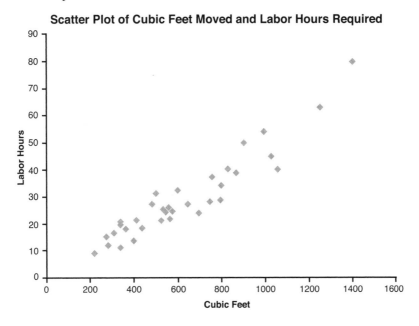

INTERPRETATION   Scatter plots help reveal patterns in the relationship between two numerical variables. The scatter plot for these data reveals a strong positive linear (straight line) relationship between the number of cubic feet moved and the number of labor hours required. Based on this relationship, you can conclude that the number of cubic feet being moved in a specific job is a useful predictor of the number of labor hours that are needed. Using one numerical variable to predict the value of another is more fully discussed in Chapter 10, "Simple Linear Regression."

spreadsheet solution

Scatter Plots

Chapter 2 Scatter Plot contains the scatter plot for the moving jobs data. Experiment with this scatter plot by typing your own data values in column A, rows 2 through 37.

Spreadsheet Tip CT7 (in Appendix D) discusses how you can create scatter plots from scratch.

2.3 Misusing Charts

All the charts presented in this chapter enhance our understanding of the data being presented. Such graphs are considered "good" graphs. Unfortunately, many charts that you encounter in the mass media or in formal reports are "bad" charts that misuse the techniques discussed in this chapter. Bad charts mislead, unnecessarily complicate things, or are just plain incorrect and should always be avoided.

INTERPRETATION   Using pictorial symbols, instead of bars or pies, always obscures the data and can create a false impression in the mind of the reader, especially if the pictorial symbols are representations of three-dimensional objects. In Example 1, the wine glass symbol fails to communicate that the 1997 data (6.77 million gallons) is almost twice the 1995 data (3.67 million gallons), nor does it accurately reflect that the 1992 data (2.25 million gallons) is a bit more than twice the 1.04 million gallons for 1989.

EXAMPLE 1: Australian Wine Exports to the United States.

We're drinking more. . .
Australian wine exports to the U.S.
in millions of gallons

| 1989 | 1992 | 1995 | 1997 |

Example 2 combines the inaccuracy of using a picture (grape vine) instead of a standard shape with the error of having unlabeled and improperly scaled axes. A missing *X* axis prevents the reader from immediately seeing that the 1997–1998 value is misplaced. By the scale of the graph, that data point should be closer to the rest of the data. A missing Y axis prevents the reader from getting a better sense of the rate of change in land planted through the years. Other problems also exist. Can you spot at least one more? (Hint: Compare the 1949–1950 data to the 1969–1970 data.)

EXAMPLE 2: Amount of Land Planted with Grapes for the Wine Industry.

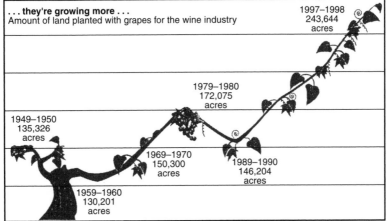

When producing your own charts, use these guidelines:

- Always choose the simplest chart that can present your data.
- Always supply a title.
- Always label every axis.
- Avoid unnecessary decorations or illustrations around the borders or in the background.
- Avoid the use of fancy pictorial symbols to represent data values.
- In two-dimensional charts, always include a scale for each axis.
- When charting non-negative values, the scale on the vertical axis should begin at zero.

One-Minute Summary

To choose an appropriate table or chart type, begin by determining whether your data are categorical or numerical.

If your data are categorical:

- Determine whether you are presenting one or two variables.
- If one variable, use a summary table and/or bar chart, pie chart, or Pareto chart.

- If two variables, use a two-way cross-classification table.

If your data are numerical:

- Determine whether you have one or two variables to present.
- If one variable, use a frequency and percentage distribution or histogram.
- If two variables, determine whether the time order of the data is important.
  - If yes, use a time-series plot.
  - If no, use a scatter plot.

Test Yourself
Short Answers

1. Which of the following graphical presentations is not appropriate for categorical data?
   (a) Pareto chart
   (b) scatter plot
   (c) bar chart
   (d) pie chart

2. Which of the following graphical presentations is not appropriate for numerical data?
   (a) histogram
   (b) pie chart
   (c) time-series plot
   (d) scatter plot

3. A type of histogram in which the categories are plotted in the descending rank order of the magnitude of their frequencies is called a:
   (a) bar chart
   (b) pie chart
   (c) scatter plot
   (d) Pareto chart

4. One of the advantages of a pie chart is that it shows that the total of all the categories of the pie adds to 100%.
   (a) True
   (b) False

5. The basic principle behind the _____ is the capability to separate the vital few categories from the trivial many categories.
   (a) scatter plot
   (b) bar chart
   (c) Pareto chart
   (d) pie chart

6. When studying the simultaneous responses to two categorical variables, you should construct a:
   (a) histogram
   (b) pie chart
   (c) scatter plot
   (d) cross-classification table

7. In a cross-classification table, the number of rows and columns:
   (a) must always be the same
   (b) must always be 2
   (c) must add to 100%
   (d) None of the above

Answer True or False:

8. Histograms are used for numerical data, whereas bar charts are suitable for categorical data.

9. A website monitors customer complaints and organizes these complaints into six distinct categories. Over the past year, the company has received 534 complaints. One possible graphical method for representing these data is a Pareto chart.

10. A website monitors customer complaints and organizes these complaints into six distinct categories. Over the past year, the company has received 534 complaints. One possible graphical method for representing these data is a scatter plot.

11. A computer company collected information on the age of its customers. The youngest customer was 12, and the oldest was 72. To study the distribution of the age of its customers, the company should use a pie chart.

12. A computer company collected information on the age of its customers. The youngest customer was 12, and the oldest was 72. To study the distribution of the age of its customers, the company can use a histogram.

13. A financial services company wants to collect information on the weekly number of transactions. To study the weekly transactions, it can use a pie chart.

14. A financial services company wants to collect information on the weekly number of transactions. To study the weekly transactions, it can use a time-series plot.

15. A professor wants to study the relationship between the number of hours a student studied for an exam and the exam score achieved. The professor can use a time-series plot.

16. A professor wants to study the relationship between the number of hours a student studied for an exam and the exam score achieved. The professor can use a bar chart.

17. A professor wants to study the relationship between the number of hours a student studied for an exam and the exam score achieved. The professor can use a scatter plot.

18. If you wanted to compare the percentage of items that are in a particular category as compared to other categories, you should use a pie chart, not a bar chart.

Fill in the blank:

19. To evaluate two categorical variables at the same time, a _____ should be developed.

20. A _____ is a vertical bar chart in which the rectangular bars are constructed at the boundaries of each class interval.

21. A _____ chart should be used when you are primarily concerned with the percentage of the total that is in each category.

22. A _____ chart should be used when you are primarily concerned with comparing the percentages in different categories.

23. A _____ should be used when you are studying a pattern between two numerical variables.

24. A _____ should be used to study the distribution of a numerical variable.

25. You have measured your pulse rate daily for 30 days. A _____ plot should be used to study the pulse rate for the 30 days.

26. You have collected data from your friends concerning their favorite soft drink. You should use a _____ chart to study the favorite soft drink of your friends.

27. You have collected data from your friends concerning the time it takes to get ready to leave their house in the morning. You should use a _____ to study this variable.

Answers to Test Yourself Short Answers

1. b
2. b
3. d
4. a
5. c
6. d
7. d
8. True

9. True

10. False

11. False

12. True

13. False

14. True

15. False

16. False

17. True

18. False

19. cross-classification table

20. histogram

21. pie chart

22. bar chart

23. scatter plot

24. histogram

25. time-series plot

26. bar chart, pie chart, or Pareto chart

27. histogram

Problems

1. An article (K. Delaney, "How Search Engine Rules Cause Sites to Go Missing," *The Wall Street Journal*, March 13, 2007, p. B1, B4) discussed the amount of Internet search results that Web surfers typically scan before selecting one. The following table represents the results for a sample of 2,369 people:

Amount of Internet Search Results Scanned	Percentage (%)
A few search results	23
First page of search results	39
First three pages	9
First two pages	19
More than first three pages	10

(a) Construct a bar chart and a pie chart.

(b) Which graphical method do you think is best to portray these data?

(c) What conclusions can you reach concerning how people scan Internet search results?

2. Medication errors are a serious problem in hospitals. The following data represent the root causes of pharmacy errors at a hospital during a recent time period:

Reason for Failure	Frequency
Additional instructions	16
Dose	23
Drug	14

Reason for Failure	Frequency
Duplicate order entry	22
Frequency	47
Omission	21
Order not discontinued when received	12
Order not received	52
Patient	5
Route	4
Other	8

(a) Construct a Pareto chart.

(b) Discuss the "vital few" and "trivial many" reasons for the root causes of pharmacy errors.

3. The following data represent the viscosity (friction, as in automobile oil) taken from 120 manufacturing batches (ordered from lowest viscosity to highest viscosity).

Chemical

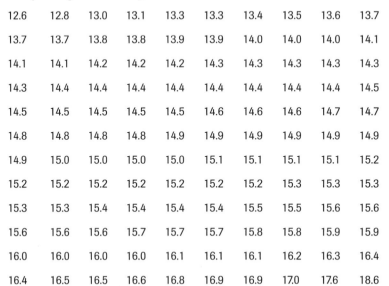

12.6	12.8	13.0	13.1	13.3	13.3	13.4	13.5	13.6	13.7
13.7	13.7	13.8	13.8	13.9	13.9	14.0	14.0	14.0	14.1
14.1	14.1	14.2	14.2	14.2	14.3	14.3	14.3	14.3	14.3
14.3	14.4	14.4	14.4	14.4	14.4	14.4	14.4	14.4	14.5
14.5	14.5	14.5	14.5	14.5	14.6	14.6	14.6	14.7	14.7
14.8	14.8	14.8	14.8	14.9	14.9	14.9	14.9	14.9	14.9
14.9	15.0	15.0	15.0	15.0	15.1	15.1	15.1	15.1	15.2
15.2	15.2	15.2	15.2	15.2	15.2	15.2	15.3	15.3	15.3
15.3	15.3	15.4	15.4	15.4	15.4	15.5	15.5	15.6	15.6
15.6	15.6	15.6	15.7	15.7	15.7	15.8	15.8	15.9	15.9
16.0	16.0	16.0	16.0	16.1	16.1	16.1	16.2	16.3	16.4
16.4	16.5	16.5	16.6	16.8	16.9	16.9	17.0	17.6	18.6

(a) Construct a frequency distribution and a percentage distribution.

(b) Construct a histogram.

(c) What conclusions can you reach about the viscosity of the chemical?

PropertyTaxes

4. The **PropertyTaxes** file contains the property taxes per capita for the 50 states and the District of Columbia.
   (a) Construct a histogram.
   (b) What conclusions can you reach concerning the property taxes per capita?

5. The following table shows the mean number of TV channels that the U.S. home received from 1985 to 2005:

TVChannels

Year	Mean Number of TV Channels Received
1985	18.8
1990	33.2
1995	41.1
2000	61.4
2005	96.4

Source: Data extracted from "At Home With More TV Channels," *USA Today*, April 10, 2007, p. A1.

   (a) Construct a time-series plot for the mean number of TV channels that the U.S. home received from 1985 to 2005.
   (b) What pattern, if any, is present in the data?
   (c) If you had to make a prediction of the mean number of TV channels that the U.S. home will receive in 2010, what would you predict?

6. The following data shows the cost of an apartment ($) and the cost of a fast-food hamburger meal ($) for 10 cities around the world.

CostofLiving

City	Rent	Hamburger
Tokyo	4,536	5.99
London	3,019	7.62
New York	3,500	5.75
Sydney	1,381	4.45
Chicago	2,300	4.99
San Francisco	2,100	5.29
Boston	1,750	4.39
Atlanta	1,250	3.70
Toronto	1,383	4.62
Rio de Janeiro	1,366	2.99

Source: Data extracted from K. Spors, "Keeping Up with…Yourself," *The Wall Street Journal*, April 11, 2005, p. R4.

(a) Construct a scatter plot.

(b) What conclusion can you reach about the relationship between apartment rent and the cost of a hamburger in these 10 cities?

Answers to Test Yourself Problems

1. (b) If you are more interested in determining which amount of searching is done most often, then the bar chart is preferred. If you are more interested in seeing the distribution of the entire set of categories, the pie chart is preferred.

   (c) Most people scan a few search results or the first page of results. Very few people scan more than the first two pages of results.

2. (b) The most important categories of medication errors are orders not received and frequency followed by dose, duplicate order entry, and omission.

3. (b)

Viscosity	Frequency	Percentage
12.0 to under 13	2	1.67%
13.0 to under 14	14	11.67%
14.0 to under 15	45	37.50%
15.0 to under 16	39	32.50%
16.0 to under 17	17	14.17%
17.0 to under 18	2	1.67%
18.0 to under 19	1	0.83%

   (c) Most of the viscosities are between 14.0 and 16.0 and very few are below 13.0 or above 17.0.

4. (b) The property taxes per capita appear to be concentrated in the middle with the center at about $1,000. Approximately 90% fall between $399 and $1,700, and the remaining 10% fall between $1,700 and $2,100.

5. (b) An obvious upward trend exists in the mean number of TV channels that the U.S. home received from 1985 to 2005.

   (c) With extrapolation, you would predict the mean number of TV channels that the U.S. home will receive in 2010 to be around 140.

6. (b) An increasing relationship seems to exist. Cities which have a higher apartment rent also have a higher cost of a hamburger meal.

References

1. Beninger, J. M., and D. L. Robyn. 1978. "Quantitative Graphics in Statistics." *The American Statistician* 32: 1–11.

2. Berenson, M. L., D. M. Levine, and T. C. Krehbiel. *Basic Business Statistics: Concepts and Applications, Eleventh Edition.* Upper Saddle River, NJ: Prentice Hall, 2009.

3. D. M. Levine. *Statistics for Six Sigma for Green Belts and Champions USING Minitab and JMP,* Upper Saddle River, NJ: Financial Times – Prentice Hall, 2006.

4. Levine, D. M., T. C. Krehbiel, and M. L. Berenson. *Business Statistics: A First Course, Fifth Edition.* Upper Saddle River, NJ: Prentice Hall, 2010.

5. Levine, D. M., D. Stephan, T. C. Krehbiel, and M. L. Berenson. *Statistics for Managers Using Microsoft Excel, Fifth Edition.* Upper Saddle River, NJ: Prentice Hall, 2008.

6. Levine, D. M., P. P. Ramsey, and R. K. Smidt, *Applied Statistics for Engineers and Scientists USING Microsoft Excel and Minitab.* Upper Saddle River, NJ: Prentice Hall, 2001.

7. Microsoft Excel 2007. Redmond, WA: Microsoft Corporation, 2006.

CHAPTER

3

Descriptive Statistics

3.1   Measures of Central Tendency
3.2   Measures of Position
3.3   Measures of Variation
3.4   Shape of Distributions
       Important Equations
       One-Minute Summary
       Test Yourself

When summarizing and describing numerical variables you need to do more than just prepare the tables and charts discussed in Chapter 2, "Presenting Data in Charts and Tables." In reading this chapter, you can learn some of the descriptive measures that identify the properties of central tendency, variation, and shape.

3.1   Measures of Central Tendency

Because the data values of most numerical variables show a tendency to group around a specific value, statisticians use a set of methods, collectively known as **measures of central tendency**, to help identify the properties of such variables. Three commonly used measures are the arithmetic mean, also known simply as the mean or average, the median, and the mode. You can calculate these measures as either sample statistics or population parameters.

The Mean

CONCEPT   A number equal to the sum of the data values, divided by the number of data values that were summed.

EXAMPLES   Many common sports statistics such as baseball batting averages and basketball points per game, mean SAT score for incoming freshmen at a college, mean age of the workers in a company, mean waiting times at a bank.

INTERPRETATION   The mean represents one way of finding the most typical value in a set of data values. As the only measure of central tendency that uses all the data values in a sample or population, the mean has one great weakness: Individual extreme values can distort the most typical value, as WORKED-OUT PROBLEM 2 illustrates.

WORKED-OUT PROBLEM 1   Although many people sometimes find themselves running late as they get ready to go to work, few measure the actual time it takes to get ready in the morning. Suppose you want to determine the typical time that elapses between your alarm clock's programmed wake-up time and the time you leave your home for work. You decide to measure actual times (in minutes) for ten consecutive working days and record the following times:

Times

Day	1	2	3	4	5	6	7	8	9	10
Time	39	29	43	52	39	44	40	31	44	35

To compute the mean time, first compute the sum of all the data values: 39 + 29 + 43 + 52 + 39 + 44 + 40 + 31 + 44 + 35, which is 396. Then, take this sum of 396 and divide by 10, the number of data values. The result, 39.6 minutes, is the mean time to get ready.

WORKED-OUT PROBLEM 2   Consider the same problem but imagine that on day 4 an exceptional occurrence such as oversleeping caused you to leave your home 50 minutes later than you had recorded for that day. That would make the time for day 4, 102 minutes; the sum of all times, 446 minutes; and the mean (446 divided by 10), 44.6 minutes.

You can see how one extreme value has dramatically changed the mean. Instead of being a number at or near the middle of the ten get-ready times, the new mean of 44.6 minutes is greater than 9 of the 10 get-ready times. In this case, the mean fails as a measure of a typical value or "central tendency."

The Median

CONCEPT   The middle value when a set of the data values have been ordered from lowest to highest value. When the number of data values is even, no natural middle value exists and you perform a special calculation to determine the median (see the Interpretation on page 46).

equation blackboard (optional)

interested in math?

The WORKED-OUT PROBLEMS calculate the mean of a sample of get-ready times. You need three symbols to write the equation for calculating the mean:

- An uppercase italic X with a horizontal line above it, $\overline{X}$, pronounced as "X bar," that represents the number that is the mean of a sample.

- A subscripted uppercase italic X (for example, X_1) that represents one of the data values being summed. Because the problem contains ten data values, there are ten X values, the first one labeled X_1, the last one labeled X_{10}.

- A lowercase italic n, which represents the number of data values that were summed in this sample, a concept also known as the **sample size**. You pronounce n as "sample size" to avoid confusion with the symbol N that represents (and is pronounced as) the population size.

Using these symbols creates the following equation:

$$\overline{X} = \frac{X_1 + X_2 + X_3 + X_4 + X_5 + X_6 + X_7 + X_8 + X_9 + X_{10}}{n}$$

By using an ellipsis (…), you can abbreviate the equation as

$$\overline{X} = \frac{X_1 + X_2 + \cdots + X_{10}}{n}$$

Using the insight that the value of the last subscript will always be equal to the value of n, you can generalize the formula as

$$\overline{X} = \frac{X_1 + X_2 + \cdots + X_n}{n}$$

By using the uppercase Greek letter sigma, Σ, a standard symbol that is used in mathematics to represent the summing of values, you can further simplify the formula as

$$\overline{X} = \frac{\Sigma X}{n}$$

or more explicitly as

$$\overline{X} = \frac{\sum_{i=1}^{n} X_i}{n}$$

in which i represents a placeholder for a subscript and the $i = 1$ and n below and above the sigma represent the range of the subscripts used.

EXAMPLES   Economic statistics such as median household income for a region; marketing statistics such as the median age for purchasers of a consumer product; in education, the established middle point for many standardized tests.

INTERPRETATION   The median splits the set of ranked data values into two parts that have an equal number of values. Extreme values do not affect the median, making the median a good alternative to the mean when such values occur.

When an even number of data values are to be summarized, you calculate the median by taking the mean of the two values closest to the middle, when all values are ranked from lowest to highest. For example, if you have six ranked values, you calculate the mean of the third and fourth ranked values. If you have ten ranked values, you calculate the mean of the fifth and sixth ranked values.

important point ✏

When you calculate the median for a very large number of values, you might not be able to quickly identify the middle value (when the number of data values is odd) or the middle two values (when the number of data values is even). To quickly determine the middle position, add 1 to the number of data values and divide that sum by 2. The result points to which value is the median. For example, if you had 127 values, divide 128 by 2 to get 64, which means that the median is the 64th ranked value. If you had 70 values, divide 71 by 2 to get 35.5, which means that the median is the mean of the 35th and 36th ranked values.

WORKED-OUT PROBLEM 3   You need to determine the median age of a group of employees whose individual ages are 47, 23, 34, 22, and 27. You calculate the median by first ranking the ages from lowest to highest: 22, 23, 27, 34, and 47. Because you have five values, the natural middle is the third ranked value, 27, making the median 27. This means that half the workers are 27 years old or younger and half the workers are 27 years old or older.

WORKED-OUT PROBLEM 4   You need to determine the median for the original set of ten get-ready times from WORKED-OUT PROBLEM 1 on page 44 that was used to explain the mean. Ordering these values from lowest to highest, you have

Time	29	31	35	39	39	40	43	44	44	52
Ordered Position	1st	2nd	3rd	4th	5th	6th	7th	8th	9th	10th

Because an even number of data values exists (ten), you calculate the mean of the two values closest to the middle—that is, the fifth and sixth ranked values, 39 and 40. The mean of 39 and 40 is 39.5, making the median 39.5 minutes for the set of ten times to get ready.

equation blackboard (optional)

Using the *n* symbol previously defined on page 44, you can define the median as:

$$\text{Median} = \frac{n+1}{2} th \text{ ranked value}$$

interested in math?

The Mode

CONCEPT   The value (or values) in a set of data values that appears most frequently.

EXAMPLES   The most common score on an exam, the most likely income, the commuting time that occurs most often.

INTERPRETATION   Some sets of data values have no mode—all the unique values appear the same number of times. Other sets of data values can have more than one mode, such as the get-ready times on page 44 in which two modes occur, 39 minutes and 44 minutes, because each of these values appears twice and all other values appear once.

Like the median, extreme values do not affect the mode. However, unlike the median, the mode can vary much more from sample to sample than the median or the mean.

3.2 Measures of Position

Measures of position describe the relative position of a data value of a numerical variable to the other values of the variable. Statisticians often use measures of position to compare two sets of data values. Two commonly encountered measures of position are the quartile and the standard (Z) score (discussed as part of Section 3.3).

Quartiles

CONCEPT   The three values that split a set of ranked data values into four equal parts, or quartiles. The **first quartile**, Q_1, is the value such that 25.0% of the ranked data values are smaller and 75.0% are larger. The **second quartile**, Q_2, is another name for the median, which, as discussed previously, splits the ranked values into two equal parts. The **third quartile**, Q_3, is the value such that 75.0% of the ranked values are smaller and 25.0% are larger.

EXAMPLE   Standardized tests that report results in terms of quartiles.

INTERPRETATION   Quartiles help bring context to a particular value that is part of a large set of values. For example, learning that you scored 25 (out of 36) on a standardized test would not be as informative as learning that you scored in the third quartile, that is, in the top 25% of all scores.

To determine the ranked value that defines the first quartile, add 1 to the number of data values and divide that sum by 4. For example, for 11 values, add 1 to 11 to get 12 and divide 12 by 4 to get 3 and determine that the third ranked value is the first quartile. To determine the ranked value that defines the third quartile, add 1 to the number of data values, divide that sum by 4, and multiply the quotient by 3. For the example of 11 values, the ninth ranked value is the third quartile (1 + 11 is 12, 12/4 is 3, 3 × 3 is 9). To determine the ranked value that defines the second quartile, use the instructions for calculating the median given on page 44.

When the result of a quartile rank calculation is not an integer, use the following procedure:

1. Select the ranked value whose rank is immediately below the calculated rank and select the ranked value whose rank is immediately above the calculated rank. For example, if the result of a quartile rank calculation is 3.75, select the third and fourth ranked values.

2. If the two ranked values selected are the same number, then the quartile is that number. If the two values are different numbers, continue with steps 3 through 5.

3. Multiply the larger ranked value by the decimal fraction of the calculated rank. (The decimal fraction will be either 0.25, 0.50, or 0.75).

4. Multiply the smaller ranked value by 1 minus the decimal fraction of the calculated rank.

5. Add the two products to determine the quartile value.

For example, if you had ten values, the calculated rank for the first quartile would be 2.75 (10 + 1 is 11, 11/4 is 2.75). Because 2.75 is not an integer, you would select the second and third ranked values. If these two values were the same, then the first quartile would be the shared value. Otherwise, by steps 3 and 4, you multiply the third ranked value by 0.75 and multiply the second ranked value by 0.25 (1 – 0.75 is 0.25). Then, by step 5, you would add these products together to get the first quartile.

For measures such as standardized test scores, another statistic, the **percentile**, is often used in addition to the quartile. The percentile expresses the percentage of ranked values that are lower than the result being reported. By the definitions given earlier, the first quartile, Q_1, is the 25th percentile; the second quartile, Q_2, is the 50th percentile; and third quartile, Q_3, is the 75th percentile. A score reported as being in the 99th percentile would be exceptional because that score is greater than 99% of all scores; that is, the score is in the top 1% of all scores.

WORKED-OUT PROBLEM 5   You are asked to determine the first quartile for the ranked get-ready times first shown on page 44 and shown here.

Time	29	31	35	39	39	40	43	44	44	52
Ranked Value	1st	2nd	3rd	4th	5th	6th	7th	8th	9th	10th

You first add 1 to 10, the number of values, and divide by 4 to get 2.75 to identify the second and third ranked values, 31 and 35. You multiply 35, the larger value, by the decimal fraction 0.75 to get 26.25. You multiply 31, the smaller value, by the decimal fraction 0.25 (which is 1 – 0.75) to get 7.75, and add 26.25 and 7.75 to produce 34, the first quartile value, indicating that 25% of the get-ready times are 34 minutes or less and that the other 75% are 34 minutes or more.

WORKED-OUT PROBLEM 6   You are asked to determine the third quartile for the ranked get-ready times. You add 1 to 10 to get 11, divide by 4 to get 2.75, and multiply by 3 to get 8.25, and therefore, select the 8th and 9th ordered values, 44 and 44. Per step 2 on page 45, the third quartile is 44. Had the 9th value been 48, you would have multiplied 48 by 0.25 to get 12 and multiplied 44 by 0.75 to get 33 and then added 12 and 33 to get 45, the third quartile value, per steps 3 through 5 on page 45.

equation blackboard (optional)

interested in math?

Using the equation for the median developed earlier,

$$\text{Median} = \frac{n+1}{2} \text{ ranked value,}$$

you can express the **first quartile, Q_1,** as

$$Q_1 = \frac{n+1}{4} \text{ th ranked value}$$

and the **third quartile, Q_3,** as

$$Q_3 = \frac{3(n+1)}{4} \text{ th ranked value}$$

WORKED-OUT PROBLEM 7   You conduct a study that compares the cost for a restaurant meal in a major city to the cost of a similar meal in the suburbs outside the city. You collect meal cost per person data from a sample of 50 city restaurants and 50 suburban restaurants and arrange the 100 values in two ranked sets as follows:

spreadsheet solution

Measures of Central Tendency and Position

Chapter 3 Descriptive (shown below) contains examples of worksheet formulas that use the **AVERAGE**, **MEDIAN**, and **MODE** functions that calculate the mean, median, and mode of a set of ranked values. The spreadsheet also calculates the first and third quartiles using a set of formulas (discussed in Appendix E, Section E.3) and are not shown in the figure.

Spreadsheet Tip FT1 in Appendix D further explains the functions that calculate measures of central tendency.

	A	B	C	D	
1	**Descriptive Statistics (using formulas)**				
2					
3	**Data**		**Statistics**		
4	29	**Arithmetic Mean**		39.6	=AVERAGE(A4:A13)
5	31	**Median**		39.5	=MEDIAN(A4:A13)
6	35	**Mode**		39	=MODE(A4:A13)
7	39	only the first mode is reported			
8	39	**First Quartile**		34.00	=G19
9	40	**Third Quartile**		44.00	=G24
10	43				
11	44				
12	44				
13	52				

RestCost

City Cost Data

13	21	22	22	24	25	26	26	26	26
30	32	33	34	34	35	35	35	35	36
37	37	39	39	39	40	41	41	41	42
43	44	45	46	50	50	51	51	53	53
53	55	57	61	62	62	62	66	68	75

Suburban Cost Data

21	22	25	25	26	26	27	27	28	28
28	29	31	32	32	35	35	36	37	37
37	38	38	38	39	40	40	41	41	41
42	42	43	44	47	47	47	48	50	50
50	50	50	51	52	53	58	62	65	67

Due to the many data values involved, you decide to create a worksheet similar to the model in **Chapter 3 Descriptive**. Into a blank worksheet, you enter the city cost data into column A and the suburban cost data into column B. You add formulas for various descriptive measures, creating a file similar to **Chapter 3 Worked-out Problem 7**, using the row 1 cells for column labels.

From the results, you note the following:

Descriptive Statistics		
	City	Surburban
Mean	41.46	39.96
Median	39.5	39.5
Mode	26	50
only the first mode is reported		
First Quartile	32.75	30.50
Third Quartile	51.50	48.50

- The mean cost of city meals, $41.46, is higher than the mean cost of suburban meals, $39.96.

- The median cost of a city meal, $39.50, is the same as the median suburban cost, $39.50.

- The first and third quartiles for city meals ($32.75 and $51.50) are higher than for suburban meals ($30.50 and $48.50).

The first and third facts enable you to conclude that the cost of a restaurant meal per person is higher in the city than it is in the suburbs.

3.3 Measures of Variation

Measures of **variation** show the amount of **dispersion**, or spread, in the data values of a numerical variable. Four frequently used measures of variation are the range, the variance, the standard deviation, and the Z score, all of which can be calculated as either sample statistics or population parameters.

The Range

CONCEPT   The difference between the largest and smallest data values in a set of data values.

EXAMPLES   The daily high and low temperatures, the stock market 52-week high and low closing prices, the fastest and slowest times for timed sporting events.

INTERPRETATION   The range is the number that represents the largest possible difference between any two values in a set of data values. The greater the range, the greater the variation in the data values.

WORKED-OUT PROBLEM 8   For the get-ready times data first presented on page 46, the range is 23 minutes (52–29). For the restaurant meal study, for the city meal cost data, the range is $62, and the range for the suburban meal cost data is $46. You can conclude that meal costs in the city show much more variation than suburban meal costs.

For a set of data values, the range is equal to

Range = largest value – smallest value

The Variance and the Standard Deviation

CONCEPT   Two measures that tell you how a set of data values fluctuate around the mean of the variable. The standard deviation is the positive square root of the variance.

EXAMPLE   The variance among SAT scores for incoming freshmen at a college, the standard deviation of the time shoppers take in a supermarket, the standard deviation of the annual return of a certain type of mutual funds.

INTERPRETATION   The variance and standard deviation, usually accompanied by the mean, help you to know how a set of data values distributes around its mean. For almost all sets of data values, most values lie within an interval of plus and minus one standard deviation above and below the mean. Therefore, determining the mean and the standard deviation usually helps you define the range in which the majority of the data values occur.

To calculate the variance, you take the difference between each data value and the mean, square this difference and then sum the squared differences. You then take this sum of squares (or *SS*) and divide it by either 1 less than the number of data values, if you have sample data, or the number of data values, if you have population data. The result is the variance. Because calculating the variance includes squaring the difference between each value and the mean, a step that always produces a non-negative number, the variance can never be negative. And as the positive square root of such a non-negative number, the standard deviation can never be negative, either.

You might wonder about the complexity of the calculation of the variance. Consider the simplest measure of variation: taking the difference between each value and the mean and summing these differences. However, by the properties and the definition of the mean, the result of such calculations would be 0 for every set of data values, which would not be very useful in comparing one set to another!

WORKED-OUT PROBLEM 9   You want to calculate the variance and standard deviation for the get-ready times first presented on page 44. As first steps, you calculate the difference between each of the 10 individual times and the mean (39.6 minutes), square those differences, and sum the squares. (Table 3.1 shows these first steps.)

TABLE 3.1

First Steps Toward Calculating the Variance and Standard Deviation for the Get-Ready Times Data

Day	Time	Difference: Time Minus Mean (39.6)	Square of Difference
1	39	−0.6	0.36
2	29	−10.6	112.36
3	43	3.4	11.56
4	52	12.4	153.76
5	39	−0.6	0.36
6	44	4.4	19.36
7	40	0.4	0.16
8	31	−8.6	73.96
9	44	4.4	19.36
10	35	−4.6	21.16
		Sum of Squares:	412.40

Because these data are a sample of get-ready times, the sum of squares, 412.40, is divided by one less than the number of data values, 9, to get 45.82, the sample variance. The square root of 45.82 (6.77, after rounding) is the sample standard deviation. You can then reasonably conclude that most get-ready times are between 32.83 (39.6 − 6.77) minutes and 46.37 (39.6 + 6.77) minutes, a statement that inspection of the data values confirms.

WORKED-OUT PROBLEM 10   You want to determine the standard deviation for the restaurant meal study. For city meal costs, the standard deviation is $13.89, and you determine that the majority of meals will cost between $27.57 and $55.35 (the mean $41.46 ± $13.89). For suburban meal costs, the standard deviation is $11.14, and you determine that the majority of those meals will cost between $28.82 and $51.10 (the mean $39.96 ± $11.14).

calculator keys

Mean, Median, Standard Deviation, Variance

To calculate descriptive statistics for a variable, first press [2nd][STAT][▶][▶] to display the Math menu. Next, select the appropriate statistic and press [ENTER]. Then type the name of the variable for which you have previously entered the data values and press [ENTER].

For example, to calculate the mean for data entered as the values for variable L1, you select 3:mean(, press [ENTER], press [2nd][1] (to type the variable name L1) and then press [ENTER]. The value of the mean appears on a new line, and your display will be similar to this:

```
mean(L1
                39.6
```

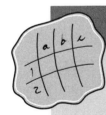

spreadsheet solution

Measures of Variation

Chapter 3 Descriptive ATP (shown here) contains measures of variation, among other descriptive statistics, for the get-ready times data as calculated by the Analysis ToolPak **Descriptive Statistics** procedure. Note that these results also include measures of central tendency, providing you with an alternative way of calculating such statistics.

	A	B
1	Get-Ready Time	
2		
3	Mean	39.6
4	Standard Error	2.140613
5	Median	39.5
6	Mode	39
7	Standard Deviation	6.769211
8	Sample Variance	45.82222
9	Kurtosis	0.13751
10	Skewness	0.085757
11	Range	23
12	Minimum	29
13	Maximum	52
14	Sum	396
15	Count	10

Spreadsheet Tip ATT2 explains how to use the **Descriptive Statistics** procedure and Tip FT2 discusses functions for measures of variation (see Appendix D).

equation blackboard (optional)

Using symbols first introduced earlier in this chapter, you can express the sample variance and the sample standard deviation as

$$\text{Sample variance} = S^2 = \frac{\Sigma(X_i - \overline{X})^2}{n-1}$$

interested in math?

$$\text{Sample standard deviation} = S = \sqrt{\frac{\Sigma(X_i - \overline{X})^2}{n-1}}$$

To calculate the variance and standard deviation for population data, change the divisor from one less than the sample size (the number of data values in the sample) to the number of data values in the population, a value known as the **population size** and represented by an italicized uppercase N.

$$\text{Population variance} = \sigma^2 = \frac{\Sigma(X_i - \mu)^2}{N}$$

$$\text{Population standard deviation} = \sigma = \sqrt{\frac{\Sigma(X_i - \mu)^2}{N}}$$

Statisticians use the lowercase Greek letter sigma, σ, to represent the population standard deviation, replacing the uppercase italicized S. In statistics, symbols for population parameters are always Greek letters. (Note that the lowercase Greek letter mu, μ, which represents the *population* mean, replaces the sample mean, $\overline{X}$, in the equations for the population variance and standard deviation.

Standard (Z) Score

CONCEPT   The number that is the difference between a data value and the mean of the variable, divided by the standard deviation.

EXAMPLE   The Z score for a particular incoming freshman's SAT score, the Z score for the get-ready time on day 4.

INTERPRETATION   Z scores help you determine whether a data value is an extreme value, or *outlier*—that is, far from the mean. As a general rule, a data value's Z score that is less than −3 or greater than +3 indicates that the data value is an extreme value.

WORKED-OUT PROBLEM 11   You need to know whether any of the times from the set of ten get-ready times (see page 44) could be considered outliers. You calculate Z scores for each of those times and compare (see Table 3.2). From these results you learn that the greatest positive Z score was 1.83 (for the day 4 value) and the greatest negative Z score was −1.27 (for the day 8 value). Because no Z score is less than −3 or greater than +3, you conclude that none of the get-ready times can be considered extreme.

TABLE 3.2

Table of Z Score Calculations for the Get-Ready Times Sample

Day	Time	Time Minus Mean	Z Score
1	39	−0.6	−0.09
2	29	−10.6	−1.57
3	43	3.4	0.50
4	52	12.4	1.83
5	39	−0.6	−0.09
6	44	4.4	0.65
7	40	0.4	0.06
8	31	−8.6	−1.27
9	44	4.4	0.65
10	35	−4.6	−0.68

equation blackboard (optional)

Using symbols presented earlier in this chapter, you can express the Z score as

$$Z \text{ score} = Z = \frac{X - \bar{X}}{S}$$

interested in math?

3.4 Shape of Distributions

Shape, a third important property of a set of numerical data, describes the pattern of the distribution of data values through the range of the data values. The shape may be symmetric, left-skewed, or right-skewed. Later in this book, you learn that determining shape often has a second purpose—some statistical methods are invalid if the set of data values is too badly skewed.

Symmetrical Shape

CONCEPT   A set of data values in which the mean equals the median value and each half of the curve is a mirror image of the other half of the curve.

EXAMPLE   Scores on a standardized exam, actual amount of soft drink in a one-liter bottle.

Left-Skewed Shape

CONCEPT   A set of data values in which the mean is less than the median value and the left tail of the distribution is longer than the right tail of the distribution. Also known as negative skew.

EXAMPLE   Scores on an exam in which most students score between 80 and 100, whereas a few students score between 10 and 79.

Right-Skewed Shape

CONCEPT   A set of data values in which the mean is greater than the median value and the right tail of the distribution is longer than the left tail of the distribution. Also known as positive skew.

EXAMPLE   Prices of new homes, annual family income.

INTERPRETATION   Right or positive skewness occurs when the data set contains some extremely high data values (that increase the mean). Left or negative skewness occurs when some extremely low values decrease the mean. The set of data values for a variable are symmetrical when low and high values balance each other out.

When identifying shape, you should avoid the common pitfall of thinking that the side of the histogram in which most data values cluster closely together is the direction of the skew. For example, consider the three histograms shown on the next page. Clustering in the first histogram appears toward the right of the histogram, but the pattern is properly labeled left-skewed. To see the shape more clearly, statisticians create area-under-the-curve or distribution graphs, in which a plotted, curved line represents the tops of all the bars. The equivalent distribution graphs for the three histograms are shown below the histograms. If you remember that in such graphs the longer tail points to the skewness, you will never wrongly identify the direction of the skew.

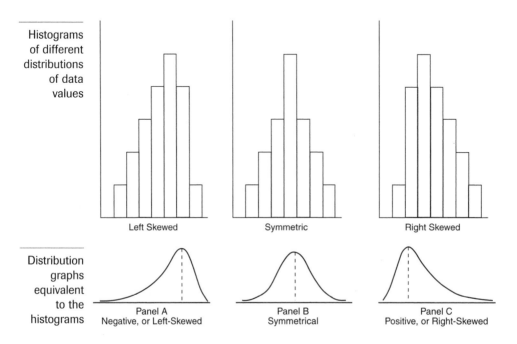

Histograms of different distributions of data values

Left Skewed   Symmetric   Right Skewed

Distribution graphs equivalent to the histograms

Panel A
Negative, or Left-Skewed

Panel B
Symmetrical

Panel C
Positive, or Right-Skewed

In lieu of graphing a distribution, a skewness statistic can also be calculated. For this statistic, a value of 0 means a perfectly symmetrical shape.

WORKED-OUT PROBLEM 12   You want to identify the shape of the NBA fan cost index data first presented on page 27. You examine the histogram of these data (see page 28) and determine that the distribution appears to be right-skewed because more low values than high values exist.

WORKED-OUT PROBLEM 13   In the figure on page 54, Microsoft Excel calculated the skewness for the get-ready times data as 0.086. Because this value is so close to 0, you conclude that the distribution of get-ready times around the mean is also approximately symmetric.

The Box-and-Whisker Plot

CONCEPT   For a set of data values, the five numbers that correspond to the smallest value, the first quartile Q_1, the median, the third quartile Q_3, and the largest value.

INTERPRETATION   The five-number summary concisely summarizes the shape of a set of data values. This plot determines the degree of symmetry (or skewness) based on the distances that separate the five numbers. To compare these distances effectively, you can create a **box-and-whisker plot**. In this plot, the five numbers are plotted as vertical lines, interconnected so as, with some imagination, to form a "box" from which a pair of cat whiskers sprout.

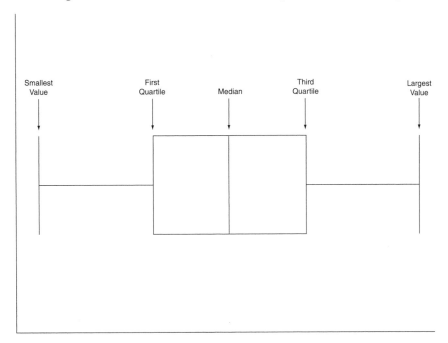

A box-and-whisker plot shows a symmetric shape for a set of data values if the following relationships are present in the plot:

- The distance from the line that represents the smallest value to the line that represents the median equals the distance from the line that represents the median to the line that represents the largest value.

- The distance from the line that represents the smallest value to the line that represents the first quartile equals the distance from the line that represents the third quartile to the line that represents the largest value.

- The distance from the line that represents the first quartile to the line that represents the median equals the distance from the line that represents the median to the line that represents the third quartile.

A box-and-whisker plot shows a right-skewed shape for a set of data values if the following relationships are present in the plot:

- The distance from the line that represents the median to the line that represents the largest value is greater than the distance from the line that represents the smallest value to the line that represents the median.

- The distance from the line that represents the third quartile to the line that represents the largest value is greater than the distance from the line that represents the smallest value to the line that represents the first quartile.

- The distance from the line that represents the first quartile to the line that represents the median is less than the distance from the line that represents the median to the line that represents the third quartile.

A box-and-whisker plot shows a left-skewed shape for a set of data values if the following relationships are present in the plot:

- The distance from the line that represents the smallest value to the line that represents the median is greater than the distance from the line that represents the median to the line that represents the largest value.

- The distance from the line that represents the smallest value to the line that represents the first quartile is greater than the distance from the line that represents the third quartile to the line that represents the largest value.

- The distance from the line that represents the first quartile to the line that represents the median is greater than the distance from the line that represents the median to the line that represents the third quartile.

WORKED-OUT PROBLEM 14   The following figure represents a box-and-whisker plot of the times to get ready in the morning:

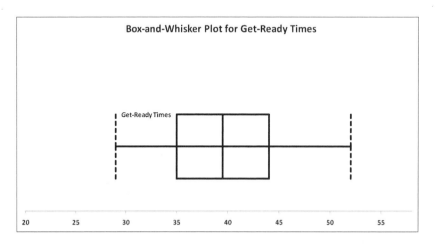

The box-and-whisker plot seems to indicate an approximately symmetric distribution of the time to get ready. The line that represents the median in the middle of the box is approximately equidistant between the ends of the box, and the length of the whiskers does not appear to be very different.

WORKED-OUT PROBLEM 15   You seek to better understand the shape of the restaurant meal cost study data used in an earlier worked-out

problem. You create box-and-whisker plots for the meal cost of both the city
and suburban groups.

Box-and-Whisker Plot for Restaurant Meals Cost Study

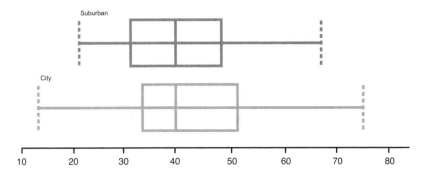

In examining the box-and-whisker plot for the city meal costs, you discover
the following:

- The distance from the line that represents the smallest value ($13) to
  the line that represents the median ($39.50) is much less than the dis-
  tance from the line that represents the median to the line that represents the highest value ($75).
- The distance from the line that represents the smallest value ($13) to
  the line that represents the first quartile ($32.75) is less than the dis-
  tance from the line that represents the third quartile ($51.50) to the
  line that represents the highest value ($75).
- The distance from the line that represents the first quartile ($32.75) to
  the line that represents the median ($39.50) is less than the distance
  from the line that represents the median ($39.50) to the line that rep-
  resents the third quartile ($51.50).

You conclude that the restaurant meal costs for city restaurants are right-
skewed.

In examining the box-and-whisker plot for the suburban meal costs, you discover the following:

- The distance from the line that represents the smallest value ($21) to
  the line that represents the median ($39.50) is much less than the dis-
  tance from the line that represents the median to the line that represents the highest value ($67).
- The distance from the line that represents the smallest value ($21) to
  the line that represents the first quartile ($30.50) is much less than the

distance from the line that represents the third quartile ($48.50) to the highest value ($67).

- The distance from the line that represents the first quartile ($30.50) to the line that represents the median ($39.5) is the same as the distance from the line that represents the median ($39.5) to the line that represents the third quartile ($48.50).

You conclude that the restaurant meal costs for suburban restaurants are right-skewed.

In comparing the city and suburban meal cost, you conclude that the city cost is higher than the suburban cost, because the minimum value, first quartile, third quartile, and maximum value are higher for the city restaurants (although the medians are the same).

calculator keys

Box-and-Whisker Plots

To display a box-and-whisker plot for a variable:

1. Press [2nd][Y=] to display the STAT PLOT menu.

2. Select 1:Plot1 and press [ENTER].

3. In the Plot1 screen, select **On** and press [ENTER]. Select the second choice of the second line of plot types (a thumbnail box-and-whisker plot) and press [ENTER]. Type the variable name as the **Xlist** value. Press [GRAPH]. (Keep **Freq** as 1.)

If you cannot see your plot, press [ZOOM] and then select **9:ZoomStat** and press [ENTER]. This re-centers your graph on the display. If you are creating a box-and-whisker plot for variable L1, your screen will look similar to this just before you press [GRAPH]:

Important Equations

Mean: (3.1) $\bar{X} = \dfrac{\sum X_i}{n}$

Median: (3.2) $\text{Median} = \dfrac{n+1}{2}th$ ranked value

First Quartile Q_1: (3.3) $Q_1 = \dfrac{n+1}{4}th$ ranked value

Third Quartile Q_3: (3.4) $Q_3 = \dfrac{3(n+1)}{4}th$ ranked value

Range: (3.5) Range = *largest value – smallest value*

Sample Variance: (3.6) $S^2 = \dfrac{\sum(X_i - \bar{X})^2}{n-1}$

Sample Standard Deviation: (3.7) $S = \sqrt{\dfrac{\sum(X_i - \bar{X})^2}{n-1}}$

Population Variance: (3.8) $\sigma^2 = \dfrac{\sum(X - \mu)^2}{N}$

Population Standard Deviation: (3.9) $\sigma = \sqrt{\dfrac{\sum(X - \mu)^2}{N}}$

Z Scores: (3.10) $Z = \dfrac{X - \bar{X}}{S}$

One-Minute Summary

The properties of central tendency, variation, and shape enable you to describe a set of data values for a numerical variable.

Numerical Descriptive Measures

- Central tendency
  Mean
  Median
  Mode

- Variation
  Range
  Variance
  Standard deviation
  Z scores

- Shape
  Five-number summary
  Box-and-whisker plot

Test Yourself
Short Answers

1. Which of the following statistics are measures of central tendency?
   (a) median
   (b) range
   (c) standard deviation
   (d) all of these
   (e) none of these

2. Which of the following statistics is not a measure of central tendency?
   (a) mean
   (b) median
   (c) mode
   (d) range

3. Which of the following statements about the median is not true?
   (a) It is less affected by extreme values than the mean.
   (b) It is a measure of central tendency.
   (c) It is equal to the range.
   (d) It is equal to the mode in bell-shaped "normal" distributions.

4. Which of the following statements about the mean is not true?
   (a) It is more affected by extreme values than the median.
   (b) It is a measure of central tendency.
   (c) It is equal to the median in skewed distributions.
   (d) It is equal to the median in symmetric distributions.

5. Which of the following measures of variability is dependent on every value in a set of data?
   (a) range
   (b) standard deviation
   (c) each of these
   (d) neither of these

6. Which of the following statistics cannot be determined from a box-and-whisker plot?
   (a) standard deviation
   (b) median
   (c) range
   (d) the first quartile

7. In a symmetric distribution:
   (a) the median equals the mean
   (b) the mean is less than the median
   (c) the mean is greater than the median
   (d) the median is less than the mode

8. The shape of a distribution is given by the:
   (a) mean
   (b) first quartile
   (c) skewness
   (d) variance

9. In a five-number summary, the following is not included:
   (a) median
   (b) third quartile
   (c) mean
   (d) minimum (smallest) value

10. In a right-skewed distribution:
    (a) the median equals the mean
    (b) the mean is less than the median
    (c) the mean is greater than the median
    (d) the median equals the mode

Answer True or False:

11. In a box-and-whisker plot, the box portion represents the data between the first and third quartile values.

12. The line drawn within the box of the box-and-whisker plot represents the mean.

Fill in the blanks:

13. The _____ is found as the middle value in a set of values placed in order from lowest to highest for an odd-sized sample of numerical data.

14. The standard deviation is a measure of _____.

15. If all the values in a data set are the same, the standard deviation will be _____.

16. A distribution that is negative-skewed is also called _____-skewed.

17. If each half of a distribution is a mirror image of the other half of the distribution, the distribution is called _____.

18. The median is a measure of _____.

19, 20, 21. The three characteristics that describe a set of numerical data are _____, _____, and _____.

For Questions 22 through 30, the number of days absent by a sample of nine students during a semester was as follows:

9   1   1   10   7   11   5   8   2

22. The mean is equal to _____.

23. The median is equal to _____.

24. The mode is equal to _____.

25. The first quartile is equal to _____.

26. The third quartile is equal to _____.

27. The range is equal to _____.

28. The variance is approximately equal to _____.

29. The standard deviation is approximately equal to _____.

30. The data are:
    (a) right-skewed
    (b) left-skewed
    (c) symmetrical

31. In a left-skewed distribution:
    (a) the median equals the mean
    (b) the mean is less than the median
    (c) the mean is greater than the median
    (d) the median equals the mode

32. Which of the statements about the standard deviation is true?
    (a) It is a measure of variation around the mean.
    (b) It is the square of the variance.
    (c) It is a measure of variation around the median.
    (d) It is a measure of central tendency.

33. The smallest possible value of the standard deviation is _____.

Answers to Test Yourself Short Answers

1. a
2. d
3. c
4. c
5. b
6. a
7. a
8. c
9. c
10. c
11. True
12. False
13. median
14. variation
15. 0
16. left

17. symmetric
18. central tendency
19. central tendency
20. variation
21. shape
22. 6
23. 7
24. 1
25. 1.5
26. 9.5
27. 10
28. 15.25
29. 3.91
30. b
31. b
32. a
33. 0

Problems

MoviePrices

1. The price for two tickets (including online service charges), a large popcorn, and two medium soft drinks at a sample of six theatre chains is as follows:
   $36.15 $31.00 $35.05 $40.25 $33.75 $43.00

   Source: Extracted from K. Kelly, "The Multiplex Under Siege," *The Wall Street Journal*, December 24-25, 2005, pp. P1, P5.

   (a) Compute the mean and median.
   (b) Compute the variance, standard deviation, and range.
   (c) Are the data skewed? If so, how?
   (d) Based on the results of (a) through (c), what conclusions can you reach concerning the cost of going to the movies?

Sushi

2. Tuna sushi was purchased from 13 Manhattan restaurants and tested for mercury. For each restaurant, the number of pieces needed to reach the maximum acceptable level of mercury, as defined by the Environmental Protection Agency, was determined to be:

8.6 2.6 1.6 5.2 7.7 4.7 6.4 6.2 3.6 4.9 9.9 3.3 4.1

Source: Data extracted from M. Burros, "High levels of Mercury Found in Tuna Sushi Sold in Manhattan," *The New York Times*, January 23, 2008, p. A1, A23.

(a) Compute the mean and median.

(b) Compute the first quartile and the third quartile.

(c) Compute the variance, standard deviation, and range.

(d) Construct a box-and-whisker plot.

(e) Are the data skewed? If so, how?

(f) Based on the results of (a) through (d), what conclusions can you reach concerning the number of pieces it would take to reach what the Environmental Protection Agency considers to be an acceptable level to be regularly consumed?

NBACost

3. As player salaries have increased, the cost of attending NBA professional basketball games has increased dramatically. The following data represents the Fan Cost Index, which is the cost of four tickets, two beers, four soft drinks, four hot dogs, two game programs, two caps, and the parking fee for one car for each of the 29 teams.

244.48 358.72 196.90 335.00 317.90 339.23 271.16 282.00 206.52 270.94

250.57 317.00 453.95 228.28 339.20 231.38 328.90 182.30 394.52 229.82

269.48 302.04 251.86 318.30 303.79 229.50 320.47 235.75 194.56

Source: Data extracted from *TeamMarketing.com*, November 2007.

(a) Compute the mean and median.

(b) Compute the first quartile and the third quartile.

(c) Compute the variance, standard deviation, and range.

(d) Construct a box-and-whisker plot.

(e) Are the data skewed? If so, how?

(f) Based on the results of (a) through (d), what conclusions can you reach concerning the Fan Cost Index of NBA games?

Chemical

4. The following data represent the viscosity (friction, as in automobile oil) taken from 120 manufacturing batches (ordered from lowest viscosity to highest viscosity).

12.6	12.8	13.0	13. 1	13.3	13.3	13.4	13.5	13.6	13.7
13.7	13.7	13.8	13.8	13.9	13.9	14.0	14.0	14.0	14.1
14.1	14.1	14.2	14.2	14.2	14.3	14.3	14.3	14.3	14.3
14.3	14.4	14.4	14.4	14.4	14.4	14.4	14.4	14.4	14.5

14.5	14.5	14.5	14.5	14.5	14.6	14.6	14.6	14.7	14.7
14.8	14.8	14.8	14.8	14.9	14.9	14.9	14.9	14.9	14.9
14.9	15.0	15.0	15.0	15.0	15.1	15.1	15.1	15.1	15.2
15.2	15.2	15.2	15.2	15.2	15.2	15.2	15.3	15.3	15.3
15.3	15.3	15.4	15.4	15.4	15.4	15.5	15.5	15.6	15.6
15.6	15.6	15.6	15.7	15.7	15.7	15.8	15.8	15.9	15.9
16.0	16.0	16.0	16.0	16.1	16.1	16.1	16.2	16.3	16.4
16.4	16.5	16.5	16.6	16.8	16.9	16.9	17.0	17.6	18.6

(a) Compute the mean and median.

(b) Compute the first quartile and the third quartile.

(c) Compute the variance, standard deviation, and range.

(d) Construct a box-and-whisker plot.

(e) Are the data skewed? If so, how?

(f) Based on the results of (a) through (d), what conclusions can you reach concerning the viscosity?

Answers to Problems

1. (a) Mean = \$36.53, median = \$35.60
   (b) Variance = \$19.27, standard deviation = \$4.39, range = \$12
   (c) The mean is slightly larger than the median, so the data are slightly right-skewed.
   (d) The mean cost is \$36.53 while the middle ranked cost is \$35.60. The average scatter of cost around the mean is \$4.39. The difference between the highest and the lowest cost is \$12.

2. (a) Mean = 5.292, median = 4.9
   (b) Q_1 = 3.45 Q_3 = 7.05
   (c) Variance = 5.806, standard deviation = 2.41, range = 8.3
   (e) The mean is slightly higher than the median. The difference between the largest value and Q_3 is 2.85, while the difference between Q_1 and the smallest value is 1.85, so the data are right-skewed.
   (f) The mean number of pieces it would take to reach what the Environmental Protection Agency considers to be an acceptable level to be regularly consumed is 5.292, whereas the number of pieces it would take to reach what the Environmental Protection Agency considers to be an acceptable level to be regularly consumed is below 4.9 in half the restaurants and above 4.9 in half the restaurants. The average scatter of the number of pieces around the mean is 2.41. The difference between the highest and the lowest number of pieces it would take to reach what the Environmental Protection Agency considers to be an acceptable level to be regularly consumed is 8.3.

3. (a) Mean = $282.90, median = $271.20
   (b) Q_1 = $230.60 Q_3 = $324.70
   (c) Variance = $4,040.20, standard deviation = $63.60, range = $271.60
   (e) The mean is slightly larger than the median. The difference between the largest value and Q_3 is $129.20, while the difference between Q_1 and the smallest value is $48.30, so the data are right-skewed.
   (f) The mean fan cost index is $282.90 whereas the fan cost index is below $271.20 for half the teams and above $271.20 for half the teams. The average scatter of the fan cost index around the mean is $63.60. The difference between the highest and the lowest fan cost index is $271.60.

4. (a) Mean = 14.977, median = 14.9
   (b) Q_1 = 14.3 Q_3 = 15.6
   (c) Variance = 1.014, standard deviation = 1.007, range = 6.0
   (e) The mean is slightly higher than the median. The difference between Q_1 and the median is 0.6 which is slightly less than the difference between the median and Q_3. The difference between the largest value and Q_3 is 3.0, while the difference between Q_1 and the smallest value is 1.7, so the data are slightly right-skewed.
   (f) The mean viscosity is 14.977 whereas the viscosity is below 14.9 for half the batches and above 14.9 for half the batches. The average scatter of the viscosity around the mean is 1.007. The difference between the highest and the lowest viscosity is 6.0.

References

1. Berenson, M. L., D. M. Levine, and T. C. Krehbiel. *Basic Business Statistics: Concepts and Applications, Eleventh Edition.* Upper Saddle River, NJ: Prentice Hall, 2009.

2. D. M. Levine. *Statistics for Six Sigma for Green Belts and Champions with Minitab and JMP,* Upper Saddle River, NJ: Prentice Hall, 2006.

3. Levine, D. M., T. C. Krehbiel, and M. L. Berenson. *Business Statistics: A First Course, Fifth Edition.* Upper Saddle River, NJ: Prentice Hall, 2010.

4. Levine, D. M., D. Stephan, T. C. Krehbiel, and M. L. Berenson. *Statistics for Managers Using Microsoft Excel, Fifth Edition.* Upper Saddle River, NJ: Prentice Hall, 2008.

5. Levine, D. M., P. P. Ramsey, and R. K. Smidt. *Applied Statistics for Engineers and Scientists Using Microsoft Excel and Minitab.* Upper Saddle River, NJ: Prentice Hall, 2001.

6. Microsoft Excel 2007. Redmond, WA: Microsoft Corporation, 2006.

CHAPTER

4

Probability

4.1   Events
4.2   More Definitions
4.3   Some Rules of Probability
4.4   Assigning Probabilities
       One-Minute Summary
       Test Yourself

You cannot properly learn the methods of inferential statistics without first understanding the basics of probability. This chapter reviews the probability concepts necessary to understand the statistical methods discussed in this book. If you are already familiar with probability, you still should take the time to skim this chapter, if only to learn the vocabulary used to refer to probability concepts in subsequent chapters.

4.1   Events

Events underlie all discussions about probability. Before you can define probability in statistical terms, you need to understand the meaning of an event.

Event

CONCEPT   An outcome of an experiment or survey.

EXAMPLES   Rolling a die and turning up six dots, an individual who votes for the incumbent candidate in an election, someone purchasing a specific brand of soft drink.

INTERPRETATION   Recall from Chapter 1 (see page 6) that performing experiments or conducting surveys are two important types of data sources. When discussing probability, many statisticians use the word *experiment* broadly to include surveys, so you can use the shorter definition "an outcome of an experiment" if you understand this broader usage of *experiment*. Likewise, as you read this chapter and encounter the word *experiment*, you should use the broader meaning.

Elementary Event

CONCEPT   An outcome that satisfies only one criterion.

EXAMPLES   A red card from a standard deck of cards, a voter who selected the Democratic candidate, a consumer who purchased a Diet Coke soft drink.

INTERPRETATION   Elementary events are distinguished from joint events, which meet two or more criteria.

Joint Event

CONCEPT   An outcome that satisfies two or more criteria.

EXAMPLES   A red ace from a standard deck of cards, a voter who voted for the Democratic candidate for president and the Democrat candidate for U.S. senator, a female who purchased a Diet Coke soft drink.

INTERPRETATION   Joint events are distinguished from elementary events which only meet one criterion.

4.2 More Definitions

Using the concept of **event**, you can define three more basic terms of probability.

Random Variable

CONCEPT   A variable whose numerical values represent the events of an experiment.

EXAMPLES   The number of cars arriving at a gas station in a one-hour period, the scores of students on a standardized exam, the preferences of consumers for different brands of automobiles.

INTERPRETATION   You use the phrase **random variable** to refer to a variable that has no data values until an experimental trial is performed or a survey question is asked and answered. Random variables are either discrete, in which the possible numerical values are a set of integers (or coded values in the case of categorical data); or continuous, in which any value is possible within a specific range.

Probability

CONCEPT   A number that represents the chance that a particular event will occur for a random variable.

EXAMPLES   Odds of winning a lottery, chance of rolling a seven when rolling two dice, likelihood of an incumbent winning reelection, percent chance of rain in a forecast.

INTERPRETATION   Probability determines the likelihood that a random variable will be assigned a specific value. Probability considers things that might occur in the future, and its forward-looking nature provides a bridge to inferential statistics.

Probabilities can be developed for an elementary event of a random variable or any group of joint events. For example, when rolling a standard six-sided die (see illustration), six possible elementary events correspond to the six faces of the die that contain either one, two, three, four, five, or six dots. "Rolling a die and turning up an even number of dots" is an example of an event formed from three elementary events (rolling a two, four, or six).

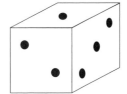

Probabilities are formally stated as decimal numbers in the range of 0 to 1. A probability of 0 indicates an event that never occurs (such an event is known as a **null event**). A probability of 1 indicates a **certain event**, an event that must occur. For example, when you roll a die, getting seven dots is a null event, because it can never happen, and getting six or fewer dots is a certain event, because you will always end up with a face that has six or fewer dots. Probabilities can also be stated informally as the "percentage chance of (something)" or as quoted odds, such as a "50-50 chance."

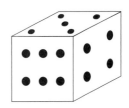

Collectively Exhaustive Events

CONCEPT   A set of events that includes all the possible events.

EXAMPLES   Heads and tails in the toss of a coin, male and female, all six faces of a die.

INTERPRETATION   When you have a set of collectively exhaustive events, one of them must occur. The coin must land on either heads or tails; the person must be male or female; the die must end with a face that has six or fewer dots. The sum of the individual probabilities associated with a set of collectively exhaustive events is always 1.

4.3 Some Rules of Probability

A set of rules govern the calculation of the probabilities of elementary and joint events.

RULE 1   The probability of an event must be between 0 and 1. The smallest possible probability value is 0. You cannot have a negative probability. The largest possible probability value is 1.0. You cannot have a probability greater than 1.0.

EXAMPLE   In the case of a die, the event of getting a face of seven has a probability of 0, because this event cannot occur. The event of getting a face with fewer than seven dots has a probability of 1.0, because it is certain that one of the elementary events of one, two, three, four, five, or six dots must occur.

RULE 2   The event that A does not occur is called A **complement** or simply **not** A, and is given the symbol A'. If $P(A)$ represents the probability of event A occurring, then $1 - P(A)$ represents the probability of event A not occurring.

EXAMPLE   In the case of a die, the complement of getting the face that contains three dots is not getting the face that contains three dots. Because the probability of getting the face containing three dots is 1/6, the probability of not getting the face that contains three dots is $(1 - 1/6) = 5/6$ or 0.833.

RULE 3   If two events A and B are **mutually exclusive**, the probability of both events A and B occurring is 0. This means that the two events cannot occur at the same time.

EXAMPLE   On a single roll of a die, you cannot get a die that has a face with three dots and also have four dots because such elementary events are mutually exclusive. Either three dots can occur or four dots can occur, but not both.

RULE 4   If two events A and B are mutually exclusive, the probability of either event A or event B occurring is the sum of their separate probabilities.

EXAMPLE   The probability of rolling a die and getting either a face with three dots or a face with four dots is 1/3 or 0.333, because these two events are mutually exclusive. Therefore, the probability is the sum of the probability of rolling a three (1/6) and the probability of rolling a four (1/6).

INTERPRETATION   You can extend this addition rule for mutually exclusive events to situations in which more than two events exist. In the case of rolling a die, the probability of turning up an even face (two, four, or six dots) is 0.50, the sum of 1/6 and 1/6 and 1/6 (3/6, or 0.50).

RULE 5   If events in a set are mutually exclusive and collectively exhaustive, the sum of their probabilities must add up to 1.0.

EXAMPLE   The events of a turning up a face with an even number of dots and turning up a face with an odd number of dots are mutually exclusive and collectively exhaustive. They are mutually exclusive, because even and odd cannot occur simultaneously on a single roll of a die. They are also collectively exhaustive, because either even or odd must occur on a particular roll. Therefore, for a single die, the probability of turning up a face with an even or odd face is the sum of the probability of turning up an even face plus the probability of turning up an odd face, or 1.0, as follows:

$$P \text{ (even or odd face)} = P \text{ (even face)} + P \text{ (odd face)}$$
$$= \frac{3}{6} + \frac{3}{6}$$
$$= \frac{6}{6} = 1$$

RULE 6   If two events A and B are not mutually exclusive, the probability of either event A or event B occurring is the sum of their separate probabilities minus the probability of their simultaneous occurrence (the joint probability).

EXAMPLE   For rolling a single die, turning up a face with an even number of dots is not mutually exclusive with turning up a face with fewer than five dots, because both events include these (two) elementary events: turning up the face with two dots and turning up the face with four dots. To determine the probability of these two events, you add the probability of having a face with an even number of dots (3/6) to the probability of having a face with fewer than five dots (4/6) and then subtract the joint probability of simultaneously having a face with an even number of dots and having a face with fewer than five dots (2/6). You can express this as follows:

$$P \text{ (even face } or \text{ face with fewer than five dots)} =$$
$$P \text{ (even face)} + P \text{ (face with fewer than five dots)} -$$
$$P \text{ (even face } and \text{ face with fewer than five dots)}$$

$$= \frac{3}{6} + \frac{4}{6} - \frac{2}{6}$$

$$= \frac{5}{6}$$

$$= 0.833$$

INTERPRETATION   This rule requires that you subtract the joint probability, because that probability has already been included twice (in the first event and in the second event). Because the joint probability has been "double counted," you must subtract it to compute the correct result.

RULE 7   If two events A and B are **independent**, the probability of both events A and B occurring is equal to the product of their individual probabilities. Two events are independent if the occurrence of one event in no way affects the probability of the second event.

EXAMPLE   When rolling a die, each roll of the die is an independent event, because no roll can affect another (although gamblers who play dice games sometimes would like to think otherwise). Therefore, to determine the probability that two consecutive rolls both turn up the face with five dots, you multiply the probability of turning up that face on roll one (1/6) by the probably of turning up that face on roll two (also 1/6). You can express this as follows:

P (face with five dots on roll one and face with five dots on roll two) =

P (face with five dots on roll one) $\times P$ (face with five dots on roll two)

$$= \frac{1}{6} \times \frac{1}{6}$$

$$= \frac{1}{36} = 0.028$$

RULE 8   If two events A and B are not independent, the probability of both events A and B occurring is the product of the probability of event A multiplied by the probability of event B occurring, given that event A has occurred.

EXAMPLE   During the taping of a television game show, contestants are randomly selected from the audience watching the show. After a particular person has been chosen, he or she does not return to the audience and cannot be chosen again, therefore making this a case in which the two events are not independent.

If the audience consists of 30 women and 20 men (50 people), what is the probability that the first two contestants chosen are male? The probability that the first contestant is male is simply 20/50 or 0.40. However, the probability that the second contestant is male is not 20/50, because when the second selection is made, the eligible audience has now only 19 males and 49 people because the first male selected cannot be selected again. Therefore, the probability that the second selection is male is 19/49 or 0.388, rounded. This means that the probability that the first two contestants are male is 0.155 as follows:

$$P \text{ (male selection first and male selection second)} =$$
$$P \text{ (male selection first)} \times P \text{ (male selection second)}$$

$$= \frac{20}{50} \times \frac{19}{49}$$
$$= \frac{380}{2,450} = 0.155$$

4.4 Assigning Probabilities

Three different approaches exist for assigning probabilities to the events of a random variable: the classical approach, the empirical approach, and the subjective approach.

Classical Approach

CONCEPT   Assigning probabilities based on prior knowledge of the process involved.

EXAMPLE   Rolling a die and assigning the probability of turning up the face with three dots.

INTERPRETATION   Classical probability often assumes that all elementary events are equally likely to occur. When this is true, the probability that a particular event will occur is defined by the number of ways the event can occur divided by the total number of elementary events. For example, when you roll a die, the probability of getting the face with three dots is 1/6 because six elementary events are associated with rolling a die. Thus you can expect that 1,000 out of 6,000 rolls of a die would turn up the face with three dots.

Empirical Approach

CONCEPT   Assigning probabilities based on frequencies obtained from empirically observed data.

EXAMPLE   Probabilities determined by polling or marketing surveys.

INTERPRETATION   The empirical approach does not use theoretical reasoning or assumed knowledge of a process to assign probabilities. Similar to the classical approach when all elementary events are equally likely, the empirical probability can be calculated by dividing the number of ways A can occur by the total number of elementary events. For example, if a poll of 500 registered voters reveals that 275 are likely to vote in the next election, you can assign the empirical probability of 0.55 (275 divided by 500).

Subjective Approach

CONCEPT   Assign probabilities based on expert opinions or other subjective methods such as "gut" feelings or hunches.

EXAMPLE   Commentators stating the odds that a political candidate will win an election or that a sports team will win a championship, a financial analyst stating the chance that a stock will increase in value by a certain amount in the next year.

INTERPRETATION   In this approach, you use your own intuition and knowledge or experience to judge the likeliest outcomes. You use the subjective approach when either the number of elementary events or actual data are not available for the calculation of relative frequencies. Because of the subjectivity, different individuals might assign different probabilities to the same event.

One-Minute Summary

Foundation Concepts

- Rules of probability
- Assigning probabilities

Test Yourself
Short Answers

1. If two events are collectively exhaustive, what is the probability that one or the other occurs?
    (a) 0
    (b) 0.50
    (c) 1.00
    (d) Cannot be determined from the information given

2. If two events are collectively exhaustive, what is the probability that both occur at the same time?

   (a) 0

   (b) 0.50

   (c) 1.00

   (d) Cannot be determined from the information given

3. If two events are mutually exclusive, what is the probability that both occur at the same time?

   (a) 0

   (b) 0.50

   (c) 1.00

   (d) Cannot be determined from the information given

4. If the outcome of event A is not affected by event B, then events A and B are said to be:

   (a) mutually exclusive

   (b) independent

   (c) collectively exhaustive

   (d) dependent

Use the following problem description when answering Questions 5 through 9:

A survey is taken among customers of a fast-food restaurant to determine preference for hamburger or chicken. Of 200 respondents selected, 125 were male and 75 were female. 120 preferred hamburger and 80 preferred chicken. Of the males, 85 preferred hamburger.

5. The probability that a randomly selected individual is a male is:

   (a) 125/200

   (b) 75/200

   (c) 120/200

   (d) 200/200

6. The probability that a randomly selected individual prefers hamburger or chicken is:

   (a) 0/200

   (b) 125/200

   (c) 75/200

   (d) 200/200

7. Suppose that two individuals are randomly selected. The probability that both prefer hamburger is:

   (a) (120/200)(120/200)

   (b) (120/200)

(c) (120/200)(119/199)

(d) (85/200)

8. The probability that a randomly selected individual prefers hamburger is:

   (a) 0/200

   (b) 120/200

   (c) 75/200

   (d) 200/200

9. The probability that a randomly selected individual prefers hamburger *or* is a male is:

   (a) 0/200

   (b) 125/200

   (c) 160/200

   (d) 200/200

10. The smallest possible value for a probability is _____.

11. The largest possible value for a probability is _____.

12. If two events are _____, they cannot occur at the same time.

13. If two events are _____, the probability that both events occur is the product of their individual probabilities.

14. In the _____ probability approach, probabilities are based on frequencies obtained from surveys.

15. In the _____ probability approach, probabilities can vary depending on the individual assigning them.

Answers to Test Yourself Short Answers

1. c

2. d

3. a

4. b

5. a

6. d

7. c

8. b

9. c

10. 0

11. 1

12. mutually exclusive

13. independent

14. empirical

15. subjective

Problems

1. Where people turn to for news is different for various age groups. Suppose that a study conducted on this issue was based on 200 respondents who were between the ages of 36 and 50, and 200 respondents

who were over age 50. Of the 200 respondents who were between the ages of 36 and 50, 82 got their news primarily from newspapers. Of the 200 respondents who were over age 50, 104 got their news primarily from newspapers. If a respondent is selected at random, what is the probability that he or she

(a) got news primarily from newspapers?

(b) got news primarily from newspapers *and* is over 50 years old?

(c) got news primarily from newspapers *or* is over 50 years old?

(d) Explain the difference in the results in (b) and (c).

(e) Suppose that two respondents are selected. What is the probability that both got news primarily from newspapers?

2. A survey of 500 men and 500 women designed to study financial tensions between couples asked how likely each was to hide purchases of clothing from his or her partner. The results were as follows:

Likely to Hide Purchase of Clothing	Men	Women	Total
Yes	62	116	178
No	438	384	822
Total	500	500	1,000

Source: Extracted from L. Wei, "Your Money Manager as Financial Therapist," *The Wall Street Journal*, November 5–6, 2005, p. B4.

If a respondent is chosen at random, what is the probability that

(a) he or she is likely to hide clothing purchases?

(b) the person is a female *and* is likely to hide clothing purchases?

(c) the person is a female *or* is likely to hide clothing purchases?

3. In 37 of the 58 years from 1950 through 2007, the S&P 500 finished higher after the first five days of trading. In 32 of those 37 years, the S&P 500 finished higher for the year. Is a good first week a good omen for the upcoming year? The following table gives the first-week and annual performance over this 58-year period:

S&P 500's Annual Performance

First five days	Year Higher	Lower
Higher	32	5
Lower	11	10

If a year is selected at random, what is the probability that
  (a) the S&P 500 finished higher for the year?
  (b) the S&P 500 finished higher after the first five days of trading?
  (c) the S&P 500 finished higher after the first five days of trading *and* the S&P 500 finished higher for the year?
  (d) the S&P 500 finished higher after the first five days of trading *or* the S&P 500 finished higher for the year?
  (e) Given that the S&P 500 finished higher after the first five days of trading, what is the probability that it finished higher for the year?

Answers to Test Yourself Problems

1. (a) 186/400 = 0.465
   (b) 104/400 = 0.26
   (c) 282/400 = 0.705
   (d) Answer (b) represents the joint probability whereas (c) represents the probability of getting news primarily from newspapers or being over 50 years old
   (e) (186/400)(185/399) = 34,410/159,600 = 0.2156

2. (a) 178/1,000 = 0.178
   (b) 116/1,000 = 0.116
   (c) 562/1,000 = 0.562

3. (a) 43/58 = 0.7414
   (b) 37/58 = 0.6379
   (c) 32/58 = 0.5517
   (d) 48/58 = 0.8276
   (e) 32/37 = 0.8649

References

1. Berenson, M. L., D. M. Levine, and T. C. Krehbiel. *Basic Business Statistics: Concepts and Applications, Eleventh Edition.* Upper Saddle River, NJ: Prentice Hall, 2009.

2. Levine, D. M., T. C. Krehbiel, and M. L. Berenson. *Business Statistics: A First Course, Fifth Edition.* Upper Saddle River, NJ: Prentice Hall, 2010.

3. Levine, D. M., D. Stephan, T. C. Krehbiel, and M. L. Berenson. *Statistics for Managers Using Microsoft Excel, Fifth Edition.* Upper Saddle River, NJ: Prentice Hall, 2008.

4. Levine, D. M., P. P. Ramsey, and R. K. Smidt. *Applied Statistics for Engineers and Scientists Using Microsoft Excel and Minitab.* Upper Saddle River, NJ: Prentice Hall, 2001.

Probability Distributions

5.1   Probability Distributions for Discrete Variables

5.2   The Binomial and Poisson Probability Distributions

5.3   Continuous Probability Distributions and the Normal Distribution

5.4   The Normal Probability Plot
      Important Equations
      One-Minute Summary
      Test Yourself

In Chapter 4, "Probability," you learned to use the rules of probability to calculate the chance that a particular event would occur. In many situations, you can use specific probability models to estimate the probability that particular events would occur.

5.1 Probability Distributions for Discrete Variables

A probability distribution for a random variable summarizes or models the probabilities associated with the events for that random variable. The distribution takes a different form depending on whether the random variable is discrete or continuous.

This section reviews probability distributions for discrete random variables and statistics related to these distributions.

Discrete Probability Distribution

CONCEPT   A listing of all possible distinct (elementary) events for a random variable and their probabilities of occurrence.

EXAMPLE   See WORKED-OUT PROBLEM 1.

INTERPRETATION   In a probability distribution, the sum of the probabilities of all the events always equals 1. This is a way of saying that the (elementary) events listed are always collectively exhaustive; that is, that one of them must occur. Although you can use a table of outcomes to develop a probability distribution (see WORKED-OUT PROBLEM 2 on page 85), you can also calculate probabilities for certain types of random variables by using a formula that mathematically models the distribution.

WORKED-OUT PROBLEM 1   You want to determine the probability of getting 0, 1, 2, or 3 heads when you toss a fair coin (one with an equal probability of a head or a tail) three times in a row. Because getting 0, 1, 2, or 3 heads represent all possible distinct outcomes, you form a table of all possible outcomes (eight) of tossing a fair coin three times as follows.

Outcome	First Toss	Second Toss	Third Toss
1	Head	Head	Head
2	Head	Head	Tail
3	Head	Tail	Head
4	Head	Tail	Tail
5	Tail	Head	Head
6	Tail	Head	Tail
7	Tail	Tail	Head
8	Tail	Tail	Tail

From this table of all eight possible outcomes, you can form the summary table shown in Table 5.1.

From this probability distribution, you can determine that the probability of tossing three heads in a row is 0.125 and that the sum of the probabilities is 1.0, as it should be for a distribution of a discrete variable.

TABLE 5.1

Probability Distribution for Tossing a Fair Coin Three Times

Number of Heads	Number of Outcomes with That Number of Heads	Probability
0	1	1/8 = 0.125
1	3	3/8 = 0.375
2	3	3/8 = 0.375
3	1	1/8 = 0.125

Another way to compute these probabilities is to extend Rule 7 on page [xref], the multiplication rule, to three events (or tosses). To get the probability of three heads, which is equal to 1/8 or 0.125 using Rule 7, you have:

$$P(H_1 \text{ and } H_2 \text{ and } H_3) = P(H_1) \times P(H_2) \times P(H_3)$$

Because the probability of heads in each toss is 0.5:

$$P(H_1 \text{ and } H_2 \text{ and } H_3) = (0.5)(0.5)(0.5)$$
$$P(H_1 \text{ and } H_2 \text{ and } H_3) = 0.125$$

The Expected Value of a Random Variable

CONCEPT   The sum of the products formed by multiplying each possible event in a discrete probability distribution by its corresponding probability.

INTERPRETATION   The expected value tells you the value of the random variable that you could expect in the "long run"; that is, after many experimental trials. The expected value of a random variable is also the mean (μ) of a random variable.

WORKED-OUT PROBLEM 2   If there are three tosses of a coin (refer to Table 5.1), you can calculate the expected value of the number of heads as shown in Table 5.2.

Expected or Mean Value = Sum of [each value × the probability of each value]

$$\text{Expected or Mean Value} = \mu = (0)(0.125) + (1)(0.375) + (2)(0.375) + (3)(0.125)$$
$$= 0 + 0.375 + 0.750 + 0.375 = 1.50$$

TABLE 5.2

Computing the Expected Value or Mean of a Probability Distribution

Number of Heads	Probability	(Number of Heads) × (Probability)
0	0.125	$(0) \times (0.125) = 0$
1	0.375	$(1) \times (0.375) = 0.375$
2	0.375	$(2) \times (0.375) = 0.75$
3	0.125	$(3) \times (0.125) = 0.125$
		Expected or Mean value = 1.50

Notice that in this example, the mean or expected value of the number of heads is 1.5, a value for the number of heads that is impossible. The mean of 1.5 heads tells you that, in the long run, if you toss three fair coins many times, the mean number of heads you can expect is 1.5.

Standard Deviation of a Random Variable (σ)

CONCEPT   The measure of variation around the expected value of a random variable. You calculate this by first multiplying the squared difference between each value and the expected value by its corresponding probability. You then sum these products and then take the square root of that sum.

WORKED-OUT PROBLEM 3   If there are three tosses of a coin (refer to Table 5.1), you can calculate the variance and standard deviation of the number of heads as shown in Table 5.3.

TABLE 5.3

Computing the Variance and Standard Deviation of a Probability Distribution

Number of Heads	Probability	(Number of Heads – Mean Number of Heads)² × (Probability)
0	0.125	$(0 - 1.5)^2 \times (0.125) = 2.25 \times (0.125) = 0.28125$
1	0.375	$(1 - 1.5)^2 \times (0.375) = 0.25 \times (0.375) = 0.09375$
2	0.375	$(2 - 1.5)^2 \times (0.375) = 0.25 \times (0.375) = 0.09375$
3	0.125	$(3 - 1.5)^2 \times (0.125) = 2.25 \times (0.125) = 0.28125$
		Total (Variance) = 0.75

σ = Square root of [Sum of (Squared differences between each value and the mean) × (Probability of the value)]

$$\sigma = \sqrt{(0-1.5)^2(0.125)+(1-1.5)^2(0.375)+(2-1.5)^2(0.375)+(3-1.5)^2(0.125)}$$
$$= \sqrt{2.25(0.125)+0.25(0.375)+0.25(0.375)+2.25(0.125)}$$
$$= \sqrt{0.75}$$

and

$$\sigma = \sqrt{0.75} = 0.866$$

INTERPRETATION   In financial analysis, you can use the standard deviation to assess the degree of risk of an investment, as WORKED-OUT PROBLEM 4 illustrates.

WORKED-OUT PROBLEM 4   Suppose that you are deciding between two alternative investments. Investment A is a mutual fund whose portfolio consists of a combination of stocks that make up the Dow Jones Industrial Average. Investment B consists of shares of a growth stock. You estimate the returns (per $1,000 investment) for each investment alternative under three economic condition events (recession, stable economy, and expanding economy), and also provide your subjective probability of the occurrence of each economic condition as follows.

Estimated Return for Two Investments Under Three Economic Conditions

		Investment	
Probability	**Economic Event**	**Dow Jones Fund (A)**	**Growth Stock (B)**
0.2	Recession	−$100	−$200
0.5	Stable economy	+100	+50
0.3	Expanding economy	+250	+350

The mean or expected return for the two investments is computed as follows:

Mean = Sum of [Each value × the probability of each value]

Mean for the Dow Jones fund = (−100)(0.2) + (100)(0.5) + (250)(0.3) = $105

Mean for the growth stock = (−200)(0.2) + (50)(0.5) + (350)(0.3) = $90

You can calculate the standard deviation for the two investments as shown in Tables 5.4 and 5.5.

TABLE 5.4

Computing the Variance and Standard Deviation for Dow Jones Fund (A)

Probability	Economic Event	Dow Jones Fund (A)	(Return – Mean Return)^2 $\times$ Probability
0.2	Recession	–$100	$(-100 - 105)^2 \times (0.2) = (42{,}025)$ $\times (0.2) = 8{,}405$
0.5	Stable economy	+ 100	$(100 - 105)^2 \times (0.5) = (25)$ $\times (0.5) = 12.5$
0.3	Expanding economy	+ 250	$(250 - 105)^2 \times (0.3) = (21{,}025)$ $\times (0.3) = 6{,}307.5$
			Total: (Variance) = 14,725

TABLE 5.5

Computing the Variance and Standard Deviation for Growth Stock (B)

Probability	Economic Event	Growth Stock (B)	(Return – Mean Return)^2 $\times$ Probability
0.2	Recession	–$200	$(-200 - 90)^2 \times (0.2) = (84{,}100)$ $\times (0.2) = 16{,}820$
0.5	Stable economy	+ 50	$(50 - 90)^2 \times (0.5) = (1{,}600)$ $\times (0.5) = 800$
0.3	Expanding economy	+ 350	$(350 - 90)^2 \times (0.3) \; (67{,}600)$ $\times (0.3) = 20{,}280$
			Total (Variance) = 37,900

σ = Square root of [Sum of (Squared differences between a value and the mean) × (Probability of the value)]

$$\sigma_A = \sqrt{(-100-105)^2(0.2)+(100-105)^2(0.5)+(250-105)^2(0.3)}$$
$$= \sqrt{14{,}725}$$
$$= \$121.35$$

$$\sigma_B = \sqrt{(-200-90)^2(0.2)+(50-90)^2(0.5)+(350-90)^2(0.3)}$$
$$= \sqrt{37{,}900}$$
$$= \$194.68$$

The Dow Jones fund has a higher mean return than the growth fund and also has a lower standard deviation, indicating less variation in the return under the different economic conditions. Having a higher mean return with less variation makes the Dow Jones fund a more desirable investment than the growth fund.

equation blackboard (optional)

interested in math?

To write the equations for the mean and standard deviation for a discrete probability distribution, you need the following symbols:

- An uppercase italic X, X, that represents a random variable.
- An uppercase italic X with an italic lowercase i subscript, X_i, that represents the ith event associated with random variable X.
- An uppercase italic N, N, that represents the number of elementary events for the random variable X. (In Chapter 3, "Descriptive Statistics," this symbol was called the population size.)

  The symbol $P(X_i)$, which represents the probability of the event X_i.

  The population mean, μ.

  The population standard deviation σ.

Using these symbols creates these equations:

The mean of a probability distribution:

$$\mu = \sum_{i=1}^{N} X_i P(X_i)$$

The standard deviation of a probability distribution:

$$\sigma = \sqrt{\sum_{i=1}^{N} (X_i - \mu)^2 P(X_i)}$$

5.2 The Binomial and Poisson Probability Distributions

As mentioned in the previous section, probability distributions for certain types of discrete random variables can be modeled using a mathematical formula. This section looks at two important discrete distributions that are widely used to compute probabilities. The first probability distribution, the binomial, is used for random variables that have only two mutually exclusive events. The second probability distribution, the Poisson, is used when you are counting the number of outcomes that occur in a unit.

The Binomial Distribution

CONCEPT   The probability distribution for a discrete random variable that meets these criteria:

- The random variable is for a sample that consists of a fixed number of experimental trials.
- The random variable has only two mutually exclusive and collectively exhaustive events, typically labeled as success and failure.
- The probability of an event being classified as a success, p, and the probability of an event being classified as a failure, $1 - p$, are both constant in all experimental trials.
- The event (success or failure) of any single experimental trial is independent of (not influenced by) the event of any other trial.

EXAMPLE   The coin tossing experiment described in WORKED-OUT PROBLEM 1 on page 84.

INTERPRETATION   Using the binomial distribution avoids having to develop the probability distribution by using a table of outcomes and applying the multiplication rule, as was done in Section 4.3. This distribution also does not require that the probability of success is 0.5, thereby allowing you to use it in more situations than the method discussed in Section 5.1.

You typically determine binomial probabilities by either using the formula in the EQUATION BLACKBOARD on page 89, by using a table of binomial probabilities, or by using software functions that create customized tables (see the figure on page 91).

When the probability of success is 0.5, you can still use the table and multiplication rule method as was done in Table 5.1. Observe from the results of that table—that the probability of zero heads is 0.125, the probability of one head is 0.375, the probability of two heads is 0.375, and the probability of three heads is 0.125—agree with the following spreadsheet results.

Binomial distributions can be symmetrical or skewed. Whenever $p = 0.5$, the binomial distribution will be symmetrical regardless of how large or small the value of the sample size, n. However, when $p \neq 0.5$, the distribution will be skewed. If $p < 0.5$, the distribution will be positive or right-skewed; if $p > 0.5$, the distribution will be negative or left-skewed. The distribution will become more symmetrical as p gets close to 0.5 and as the sample size, n, gets large.

	A	B	C
1	Binomial Probabilities		
2			
3	Data		
4	Sample size	3	
5	Probability of success	0.5	
6			
7	Statistics		
8	Mean	1.5	
9	Variance	0.75	
10	Standard deviation	0.866025	
11			
12	Binomial Probabilities Table		
13		X	P(X)
14		0	0.125
15		1	0.375
16		2	0.375
17		3	0.125

For the binomial distribution, the number of experimental trials is equivalent to the term *sample size* introduced in Chapter 1, "Fundamentals of Statistics." You calculate the mean and standard deviation for a random variable that can be modeled using the binomial distribution using the sample size, the probability of success, and the probability of failure as follows.

Binomial Distribution Characteristics

Mean
The sample size (n) times the probability of success or $n \times p$, remembering that the sample size is the number of experimental trials.

Variance
The product of these three: sample size, probability of success, and probability of failure (1 – probability of success), or

$$n \times p \times (1 - p)$$

Standard deviation
The square root of the variance, or

$$\sqrt{np(1-p)}$$

equation blackboard (optional)

For the equation for the binomial distribution, use the symbols X (random variable), n (sample size), and p (probability of success) previously introduced and add these symbols:

- A lowercase italic X, x, which represents the number of successes in the sample.

- The symbol $P(X = x \mid n, p)$, which represents the probability of the value x, given sample size n and probability of success p.

interested in math?

You use these symbols to form two separate expressions. One expression represents the number of ways you can get a certain number of successes in a certain number of trials:

$$\frac{n!}{x!(n-x)!}$$

(The symbol ! means factorial, where $n! = (n)(n-1)\ldots(1)$ so that 3! equals 6, $3 \times 2 \times 1$. 1! equals 1 and 0! is defined as being equal to 1.)

The second expression represents the probability of getting a certain number of successes in a certain number of trials *in a specific order*:

$$p^x \times (1-p)^{n-x}$$

Using these expressions forms the following equation:

$$P(X = x \mid n, p) = \frac{n!}{x!(n-x)!} p^x (1-p)^{n-x}$$

As an example, the calculations for determining the binomial probability of one head in three tosses of a fair coin (that is, for a problem in which $n = 3$, $p = 0.5$, and $x = 1$) are as follows:

$$P(X = 1) \mid n = 3, p = 0.5) = \frac{3!}{1!(3-1)!} (0.5)^1 (1-0.5)^{3-1}$$

$$= \frac{3!}{1!(2)!} (0.5)^1 (1-0.5)^2$$

$$= 3(0.5)(0.25) = 0.375$$

Using symbols previously introduced, you can write the equation for the mean and standard deviation of the binomial distribution:

$$\mu = np$$

and

$$\sigma = \sqrt{np(1-p)}$$

calculator keys

Binomial Probabilities

To calculate an exact or cumulative binomial probability, press [2nd][VARS] to display the DISTR menu and then select either **A:binompdf** (or **B:binomcdf**) (choices O: and A: on older calculators) and press [Enter]. Enter the the sample size, probability of success, and optionally, the number of successes, separated by commas, and press [**Enter**]. If you do not enter a value for the number of successes, you will get a list of probabilities that you can view by using the cursor keys.

spreadsheet solution

Binomial Probabilities

Chapter 5 Binomial calculates a table of binomial probabilities based on a sample size, probability of success, and number of successes you enter. Spreadsheet Tip FT3 in Appendix D explains the spreadsheet function BINOMDIST that calculates binomial probabilities.

WORKED-OUT PROBLEM 5   An online social networking website defines success if a Web surfer stays and views its website for more than three minutes. Suppose that the probability that the surfer does stay for more than three minutes is 0.16. What is the probability that at least four (either four or five) of the next five surfers will stay for more than three minutes?

You need to sum the probabilities of four surfers staying and five surfers staying in order to determine the probabilities that at least four surfers stay.

Spreadsheet and calculator (partial) results for this example are as follows:

	A	B	C
1	Binomial Probabilities		
2			
3	Data		
4	Sample size	5	
5	Probability of success	0.16	
6			
7	Statistics		
8	Mean	0.8	
9	Variance	0.672	
10	Standard deviation	0.8198	
11			
12	Binomial Probabilities Table		
13		X	P(X)
14		0	0.4182
15		1	0.3983
16		2	0.1517
17		3	0.0289
18		4	0.0028
19		5	0.0001

```
binompdf(5,0.16
(.4182119424  .3…
```

From the spreadsheet results:

$$P(X = 4 | n = 5, p = 0.16) = 0.0028$$
$$P(X = 5 | n = 5, p = 0.16) = 0.0001$$

Therefore, the probability of four or more surfers staying and viewing the social networking website for more than three minutes is 0.0029 (which you compute by adding 0.0028 and 0.0001) or 0.29%.

important point

The Poisson Distribution

CONCEPT   The probability distribution for a discrete random variable that meets these criteria:

- You are counting the number of times a particular event occurs in a unit.
- The probability that an event occurs in a particular unit is the same for all other units.
- The number of events that occur in a unit is independent of the number of events that occur in other units.
- As the unit gets smaller, the probability that two or more events will occur in that unit approaches zero.

EXAMPLES   Number of computer network failures per day, number of surface defects per square yard of floor covering, the number of customers arriving at a bank during the 12 noon to 1 p.m. hour, the number of fleas on the body of a dog.

INTERPRETATION   To use the Poisson distribution, you define an area of opportunity, a continuous unit of area, time, or volume in which more than one event can occur. The Poisson distribution can model many random variables that count the number of defects per area of opportunity or count the number of times items are processed from a waiting line.

You determine Poisson probabilities by applying the formula in the EQUATION BLACKBOARD on page 96, by using a table of Poisson values, or by using software functions that create customized tables (see the figure below). You can calculate the mean and standard deviation for a random variable that can be modeled using the Poisson distribution as follows:

Poisson Distribution Characteristics

Mean	The population mean, λ.
Variance	The population mean, λ, that in the Poisson distribution, is equal to the variance.
Standard deviation	The square root of the variance, or $\sqrt{\lambda}$.

WORKED-OUT PROBLEM 6   You want to determine the probabilities that a specific number of customers will arrive at a bank branch in a one-minute interval during the lunch hour: Will zero customers arrive, one customer, two customers, and so on? You determine that you can use the Poisson distribution because of the following reasons:

- The random variable is a count per unit, customers per minute.
- You assume that the probability that a customer arrives during a specific one-minute interval is the same as the probability for all the other one-minute intervals.
- Each customer's arrival has no effect on (is independent of) all other arrivals.
- The probability that two or more customers will arrive in a given time period approaches zero as the time interval decreases from one minute.

Using historical data, you determine that the mean number of arrivals of customers is three per minute during the lunch hour.

You use a spreadsheet table (see the following figure) to calculate these Poisson probabilities:

	A	B	C	D	E
1	Poisson Probabilities for Customer Arrivals				
2					
3			Data		
4	Average/Expected number of successes:			3	
5					
6	Poisson Probabilities Table				
7		X	P(X)		
8		0	0.049787		
9		1	0.149361		
10		2	0.224042		
11		3	0.224042		
12		4	0.168031		
13		5	0.100819		
14		6	0.050409		
15		7	0.021604		
16		8	0.008102		
17		9	0.002701		
18		10	0.000810		
19		11	0.000221		
20		12	0.000055		
21		13	0.000013		
22		14	0.000003		
23		15	0.000001		

From the results, you observe the following:

- The probability of zero arrivals is 0.049787.
- The probability of one arrival is 0.149361.
- The probability of two arrivals is 0.224042.

Therefore, the probability of two or fewer customer arrivals per minute at the bank during the lunch hour is 0.42319, the sum of the probabilities for zero, one, and two arrivals (0.049787 + 0.149361 + 0.224042 = 0.42319).

interested in math?

For the equation for the Poisson distribution, use the symbols X (random variable), n (sample size), p (probability of success) previously introduced and add these symbols:

- A lowercase italic E, e, which represents the mathematical constant approximated by the value 2.71828.
- A lowercase Greek symbol lambda, λ, which represents the mean number of times that the event occurs per area of opportunity.
- A lowercase italic X, x, which represents the number of times the event occurs per area of opportunity.
- The symbol $P(X = x \mid \lambda)$, which represents the probability of x, given λ.

Using these symbols forms the following equation:

$$P(X = x \mid \lambda) = \frac{e^{-\lambda}\lambda^x}{x!}$$

As an example, the calculations for determining the Poisson probability of exactly two arrivals in the next minute given a mean of three arrivals per minute is as follows:

$$P(X = 2 \mid \lambda = 3) = \frac{e^{-3}(3)^2}{2!}$$

$$= \frac{(2.71828)^{-3}(3)^2}{2!}$$

$$= \frac{(0.049787)(9)}{(2)}$$

$$= 0.224042$$

calculator keys

Poisson Probabilities

To calculate an exact or cumulative Poisson probability, press
[2nd][VARS] to display the DISTR menu and select either
C:poissonpdf(or **D:poissoncdf(** (choices B: and C: on older
calculators), and press [ENTER]. Enter the mean number of
successes and the total number of successes and press [Enter].

spreadsheet solution

Poisson Probabilities

Chapter 5 Poisson calculates a table of Poisson probabilities
based on the mean number of successes you enter.
Spreadsheet Tip FT4 in Appendix D explains the spreadsheet
function POISSON that calculates POISSON probabilities.

5.3 Continuous Probability Distributions and the Normal Distribution

Probability distributions can also be developed to model continuous random
variables. The exact mathematical expression for the probability distribution for
a continuous variable involves integral calculus and is not shown in this book.

Probability Distribution for a Continuous RandomVariable

CONCEPT   The area under a curve that represents the probabilities for a
continuous random variable.

EXAMPLE   See the example for the normal distribution on page 98.

INTERPRETATION   Probability distributions for a continuous random variable differ from discrete distributions in several important ways:

important point

- An event can take on any value within the range of the random variable
  and not just an integer value.
- The probability of any specific value is zero.

- Probabilities are expressed in terms of an area under a curve that represents the continuous distribution.

One continuous distribution, the **normal distribution**, is especially important in statistics because it can model many different continuous random variables.

Normal Distribution

important point

CONCEPT   The probability distribution for a continuous random variable that meets these criteria:

- The graphed curve of the distribution is bell-shaped and symmetrical.
- The mean, median, and mode are all the same value.
- The population mean, μ, and the population standard deviation, σ, determine probabilities.
- The distribution extends from negative to positive infinity. (The distribution has an infinite range.)
- Probabilities are always cumulative and expressed as inequalities, such as $P < X$ or $P \geq X$, where X is a value for the variable.

EXAMPLE   The normal distribution appears as a bell-shaped curve as shown in the following figure.

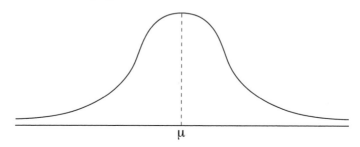

μ

INTERPRETATION   The importance of the normal distribution to statistics cannot be overstated. Probabilities associated with variables as diverse as physical characteristics such as height and weight, scores on standardized exams, and the dimension of industrial parts, tend to follow a normal distribution. Under certain circumstances, the normal distribution also approximates various discrete probability distributions such as the binomial and Poisson distributions. In addition, the normal distribution provides the basis for classical statistical inference discussed in Chapters 6 through 9.

You determine normal probabilities by using a table of normal probabilities (such as Table C.1 in Appendix C) or by using software functions. (You do not use a formula to directly determine the probabilities, because the complexities of the formula rule out its everyday use.) Normal probability tables (including Table C.1) and some software functions use a standardized normal distribution that requires you to convert an X value of a variable to its corresponding Z score (see Section 3.3). You perform this conversion by subtracting the population mean μ from the X value and dividing the resulting difference by the population standard deviation σ, expressed algebraically as follows:

$$Z = \frac{X - \mu}{\sigma}$$

When the mean is 0 and the standard deviation is 1, the X value and Z score will be the same and no conversion is necessary.

WORKED-OUT PROBLEM 7   Packages of chocolate candies have a labeled weight of 6 ounces. In order to ensure that very few packages have a weight below 6 ounces, the filling process provides a mean weight above 6 ounces. In the past, the mean weight has been 6.15 ounces with a standard deviation of 0.05 ounce. Suppose you want to determine the probability that a single package of chocolate candies will weigh between 6.15 and 6.20 ounces. To determine this probability, you use Table C.1, the table of the probabilities of the cumulative standardized normal distribution.

To use Table C.1, you must first convert the weights to Z scores by subtracting the mean and then dividing by the standard deviation, as shown here:

$$Z(\text{lower}) = \frac{6.15 - 6.15}{0.05} = 0 \qquad Z(\text{upper}) = \frac{6.20 - 6.15}{0.05} = 1.0$$

Therefore, you need to determine the probability that corresponds to the area between 0 and +1 Z units (standard deviations). To do this, you take the cumulative probability associated with 0 Z units and subtract it from the cumulative probability associated with +1 Z units. Using Table C.1, you determine that these probabilities are 0.8413 and 0.5000, respectively (see the following table).

Finding a Cumulative Area under the Normal Curve

| | Cumulative Probabilities | | | | | | | | |
Z	.00	.01	.02	.03	.04	.05	.06	.07	.08	.09
0.0	.5000	.5040	.5080	.5120	.5160	.5199	.5239	.5279	.5319	.5359
0.1	.5398	.5438	.5478	.5517	.5557	.5596	.5636	.5675	.5714	.5753
0.2	.5793	.5832	.5871	.5910	.5948	.5987	.6026	.6064	.6103	.6141
0.3	.6179	.6217	.6255	.6293	.6331	.6368	.6406	.6443	.6480	.6517
0.4	.6554	.6591	.6628	.6664	.6700	.6736	.6772	.6808	.6844	.6879
0.5	.6915	.6950	.6985	.7019	.7054	.7088	.7123	.7157	.7190	.7224
0.6	.7257	.7291	.7324	.7357	.7389	.7422	.7454	.7486	.7518	.7549
0.7	.7580	.7612	.7642	.7673	.7704	.7734	.7764	.7794	.7823	.7852
0.8	.7881	.7910	.7939	.7967	.7995	.8023	.8051	.8078	.8106	.8133
0.9	.8159	.8186	.8212	.8238	.8264	.8289	.8315	.8340	.8365	.8389
1.0	.8413	.8438	.8461	.8485	.8508	.8531	.8554	.8577	.8599	.8621

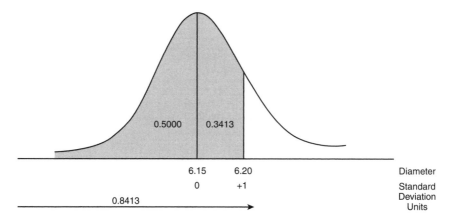

Therefore, the probability that a single package of chocolate candies will weigh between 6.15 and 6.20 ounces is 0.3413 (0.8413 – 0.5000 = 0.3413).

A spreadsheet that calculates various normal probabilities shows the same results:

	A	B	C	D	E
1	Normal Probabilities for Chocolate Candy Packages				
2					
3	Common Data				
4	Mean	6.15			
5	Standard Deviation	0.05			
6					
7	Probability for X <=			Probability for a Range	
8	X Value	6.2		From X Value	6.15
9	Z Value	1		To X Value	6.2
10	P(X<=6.2)	0.8413		Z Value for 6.15	0
11				Z Value for 6.2	1
12	Probability for X >			P(X<=6.15)	0.5000
13	X Value	6.2		P(X<=6.2)	0.8413
14	Z Value	1		P(6.15<=X<=6.2)	0.3413
15	P(X>6.2)	0.1587			

From these results, you can also see that the probability that a package weighs more than 6.20 ounces is 0.1587 (1 – 0.8413).

Using Standard Deviation Units

Because of the equivalence between Z scores and standard deviation units, probabilities of the normal distribution are often expressed as ranges of plus-or-minus standard deviation units. Such probabilities can be determined directly from Table C.1, the table of the probabilities of the cumulative standardized normal distribution.

For example, to determine the normal probability associated with the range ±3 standard deviations, you would use Table C.1 to look up the probabilities associated with $Z = -3.00$ and $Z = +3.00$:

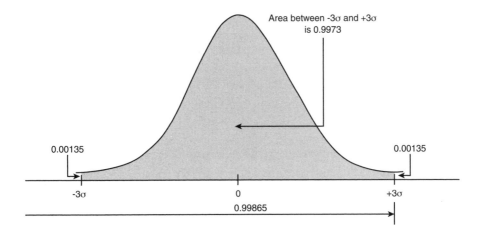

Table 5.6 represents the appropriate portion of Table C.1 for Z = −3.00. From this table excerpt, you can determine that the probability of a value less than Z = −3 units is 0.00135.

TABLE 5.6

Partial Table C.1 for Obtaining a Cumulative Area Below −3 Z Units

Z	.00	.01	.02	.03	.04	.05	.06	.07	.08	.09
.	.	.	.	.	.	.	.	.	.	.
.	.	.	.	.	.	.	.	.	.	.
.	.	.	.	.	.	.	.	.	.	.
−3.0	0.00135	0.00131	0.00126	0.00122	0.00118	0.00114	0.00111	0.00107	0.00103	0.00100

Source: Extracted from Table C.1

Table 5.7 represents the appropriate portion of Table C.1 for Z = +3.00. From this table excerpt, you can determine that the probability of a value less than Z = +3 units is 0.99865.

TABLE 5.7

Partial Table C.1 for Obtaining a Cumulative Area Below +3 Z Units

Z	.00	.01	.02	.03	.04	.05	.06	.07	.08	.09
.	.	.	.	.	.	.	.	.	.	.
.	.	.	.	.	.	.	.	.	.	.
.	.	.	.	.	.	.	.	.	.	.
+3.0	0.99865	0.99869	0.99874	0.99878	0.99882	0.99886	0.99889	0.99893	0.99897	0.99900

Source: Extracted from Table C.1

Therefore, the probability associated with the range plus-or-minus three standard deviations in a normal distribution is 0.9973 (0.99865 – 0.00135). Stated another way, the probability is 0.0027 (2.7 out of a thousand chance) that a value will not be within the range of plus-or-minus three standard deviations. Table 5.8 summarizes probabilities for several different ranges of standard deviation units.

TABLE 5.8

Probabilities for Different Standard Deviation Ranges

Standard Deviation Unit Ranges	Probability or Area Outside These Units	Probability or Area Within These Units
-1σ to $+1\sigma$	0.3174	0.6826
-2σ to $+2\sigma$	0.0455	0.9545
-3σ to $+3\sigma$	0.0027	0.9973
-6σ to $+6\sigma$	0.000000002	0.999999998

Finding the *Z* Value from the Area Under the Normal Curve

Each of the previous examples involved using the normal distribution table to find an area under the normal curve that corresponded to a specific Z value. In many circumstances you want to do the opposite of this and find the Z value that corresponds to a specific area. For example, you can find the Z value that corresponds to a cumulative area of 1%, 5%, 95%, or 99%. You can also find lower and upper Z values between which 95% of the area under the curve is contained.

To find the Z value that corresponds to a cumulative area, you locate the cumulative area in the body of the normal table, or the closest value to the cumulative area you want to find, and then determine the Z value that corresponds to this cumulative area.

WORKED-OUT PROBLEM 8   You want to find the Z values such that 95% of the normal curve is contained between a lower Z value and an upper Z value with 2.5% below the lower Z value, and 2.5% above the upper Z value. Using the following figure, you determine that you need to find the Z value that corresponds to a cumulative area of 0.025 and the Z value that corresponds to a cumulative area of 0.975.

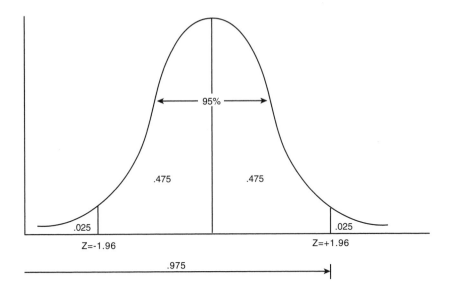

Table 5.9 contains a portion of Table C.1 that is needed to find the Z value that corresponds to a cumulative area of 0.025. Table 5.10 contains a portion of Table C.1 that is needed to find the Z value that corresponds to a cumulative area of 0.975.

TABLE 5.9

Partial Table C.1 for Finding Z Value That Corresponds to a Cumulative Area of 0.025

Z	.00	.01	.02	.03	.04	.05	.06	.07	.08	.09
.	.	.	.	.	.	.	.	.	.	.
.	.	.	.	.	.	.	.	.	.	.
-2.0	0.0228	0.0222	0.0217	0.0212	0.0207	0.0202	0.0197	0.0192	0.0188	0.0183
-1.9	0.0287	0.0281	0.0274	0.0268	0.0262	0.0256	0.0250	0.0244	0.0239	0.0233

TABLE 5.10

Partial Table C.1 for Finding Z Value That Corresponds to a Cumulative Area of 0.975

Z	.00	.01	.02	.03	.04	.05	.06	.07	.08	.09
.	.	.	.	.	.	.	.	.	.	.
.	.	.	.	.	.	.	.	.	.	.
1.9	0.9713	0.9719	0.9726	0.9732	0.9738	0.9744	0.9750	0.9756	0.9761	0.9767
2.0	0.9772	0.9778	0.9783	0.9788	0.9793	0.9798	0.9803	0.9808	0.9812	0.9817

To find the Z value that corresponds to a cumulative area of 0.025, you look in the body of Table 5.9 until you see the value of 0.025. Then you determine the row and column that this value corresponds to. Locating the value of 0.025, you see that it is located in the –1.9 row and the .06 column. Thus the Z value that corresponds to a cumulative area of 0.025 is –1.96.

To find the Z value that corresponds to a cumulative area of 0.975, you look in the body of Table 5.10 until you see the value of 0.975. Then you determine the corresponding row and column that this value belongs to. Locating the value of 0.975, you see that it is in the 1.9 row and the .06 column. Thus the Z value that corresponds to a cumulative area of 0.975 is 1.96. Taking this result along with the Z value of –1.96 for a cumulative area of 0.025 means that 95% of all the values will be between $Z = -1.96$ and $Z = 1.96$.

WORKED-OUT PROBLEM 9   You want to find the weights that will contain 95% of the packages of chocolate candy first discussed in WORKED-OUT PROBLEM 7 on page 99. In order to do so, you need to determine X in the formula

$$Z = \frac{X - \mu}{\sigma}$$

Solving this formula for X, you have

$$X = \mu + Z\sigma$$

Because the mean weight is 6.15 ounces and the standard deviation is 0.05 ounce, and 95% of the packages will be contained between –1.96 and +1.96 standard deviation (Z) units, the interval that will contain 95% of the packages will be between

$$6.15 + (-1.96)(0.05) = 6.15 - 0.098 = 6.052 \text{ ounces}$$

and

$$6.15 + (+1.96)(0.05) = 6.15 + 0.098 = 6.248 \text{ ounces}$$

spreadsheet solution

Normal Probabilities

Chapter 5 Normal calculates the normal probability for several types of problems based on the mean, standard deviation, and X value(s) you enter.

Chapter 5 Z Value determines the Z value based on the mean, standard deviation, and cumulative percentage you enter.

Spreadsheet Tip FT5 in Appendix D explains the spreadsheet functions that calculate normal probabilities.

calculator keys

Normal Probabilities

To calculate the cumulative normal probability for a specific X value: Press [2nd][VARS] to display the DISTR menu and then select **1:normalpdf** and press [Enter]. Enter the X value, the mean, and the standard deviation, separated by commas, and press [Enter].

To calculate the normal probability for a range: Press [2nd][VARS] to display the DISTR menu and then select **2:normalcdf(** and press [Enter]. Enter the lower value, the upper value, the mean, and the standard deviation, separated by commas, and press [ENTER].

To find a Z value from the area under the normal curve: Press [2nd][VARS] to display the DISTR menu and select **3:invNorm(**. Enter the area value and press [Enter].

5.4 The Normal Probability Plot

You need to determine whether a set of data values approximately follows a normal distribution in order to use many inferential statistical methods. One technique for evaluating whether the data follow a normal distribution is the **normal probability plot**.

CONCEPT   A graph that plots the relationship between ranked data values and the Z scores that these values would correspond to if the set of data values follows a normal distribution. If the data values follow a normal distribution, the graph will be linear (a straight line), as shown in the following example.

EXAMPLES

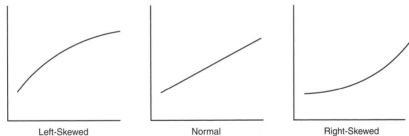

Left-Skewed   Normal   Right-Skewed

INTERPRETATION   Normal probability plots are based on the idea that the Z scores for the ranked values increase at a predictable rate for data that follow a normal distribution. The exact details to produce a normal probability plot can vary, but one common approach is called the **quantile–quantile plot**. In this method, each value (ranked from lowest to highest) is plotted on the vertical Y axis and its transformed Z score is plotted on the horizontal X axis. If the data are normally distributed, a plot of the data ranked from lowest to highest will follow a straight line. As shown in the preceding figure, if the data are left-skewed, the curve will rise more rapidly at first, and then level off. If the data are right-skewed, the data will rise more slowly at first, and then rise at a faster rate for higher values of the variable being plotted.

RestCost

WORKED-OUT PROBLEM 10   You seek to determine whether the cost of a meal at city restaurants first presented in Chapter 2, "Presenting Data in Charts and Tables," follows a normal distribution. You decide to use Microsoft Excel to produce the following normal probability plot:

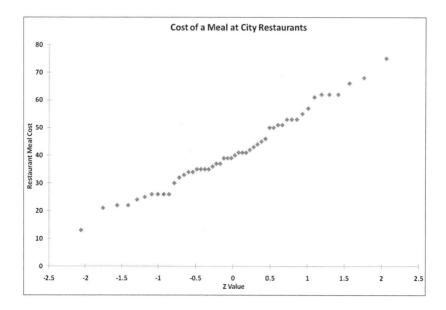

Consistent with the results of the histogram in Section 2.2, the approximate straight line that the data follow in this normal probability plot appears to indicate that the cost of a meal at city restaurants is approximately normally distributed.

Normal Probability Plots

To display a normal probability plot for a set of data values previously entered as the values of a variable, press [2nd][Y=] to display the Stat Plot menu and then select 1:Plot1 and press [ENTER]. On the Plot 1 screen, select On and press [ENTER]; select the sixth type choice (a thumbnail normal probability plot) and press [ENTER]; and then enter the name of the variable as the Data List. Press [GRAPH]. If you do not see your plot, press [ZOOM] and then select 9:ZoomStat and press [ENTER] to re-center your graph on the plot.

Important Equations

The mean of a probability distribution:

$$(5.1) \quad \mu = \sum_{t=1}^{N} X_i P(X_i)$$

The standard deviation of a discrete probability distribution:

$$(5.2) \quad \sigma = \sqrt{\sum_{i=1}^{N} (X_i - \mu)^2 P(X_i)}$$

The binomial distribution:

$$(5.3) \quad P(X = x) \mid n, p) = \frac{n!}{x!(n-x)!} p^x (1-p)^{n-x}$$

The mean of the binomial distribution:

$$(5.4) \quad \mu = np$$

The standard deviation of the binomial distribution:

$$(5.5) \quad \sigma_x = \sqrt{np(1-p)}$$

The Poisson distribution:

$$(5.6) \quad P(X = x \mid \lambda) = \frac{e^{-\lambda} \lambda^x}{x!}$$

The normal distribution: finding a Z value

$$(5.7) \quad Z = \frac{X - \mu}{\sigma}$$

The normal distribution: finding an X value

$$(5.8) \quad X = \mu + Z\sigma$$

One-Minute Summary
Probability Distributions

- Discrete probability distributions

    Expected value

    Variance σ^2 and standard deviation σ

- Is there a fixed sample size n and is each observation classified into one of two categories?

    If yes, use the binomial distribution, subject to other conditions.

    If no, use the Poisson distribution, subject to other conditions.

- Continuous probability distributions

    Normal distribution

    Normal probability plot

Test Yourself
Short Answers

1. The sum of the probabilities of all the events in a probability distribution is equal to:
    (a) 0
    (b) the mean
    (c) the standard deviation
    (d) 1

2. The largest number of possible successes in a binomial distribution is:
    (a) 0
    (b) 1
    (c) n
    (d) infinite

3. The smallest number of possible successes in a binomial distribution is:
    (a) 0
    (b) 1
    (c) n
    (d) infinite

4. Which of the following about the binomial distribution is not a true statement?
    (a) The probability of success must be constant from trial to trial.
    (b) Each outcome is independent of the other.
    (c) Each outcome may be classified as either "success" or "failure."
    (d) The random variable of interest is continuous.

5. Whenever $p = 0.5$, the binomial distribution will:
    (a) always be symmetric
    (b) be symmetric only if n is large
    (c) be right-skewed
    (d) be left-skewed

6. What type of probability distribution will the consulting firm most likely employ to analyze the insurance claims in the following problem?

    An insurance company has called a consulting firm to determine whether the company has an unusually high number of false insurance claims. It is known that the industry proportion for false claims is 6%. The consulting firm has decided to randomly and independently sample 50 of the company's insurance claims. They believe that the number of claims from the sample, 50 of which are false, will yield the information the company desires.
    (a) Binomial distribution
    (b) Poisson distribution
    (c) Normal distribution
    (d) None of the above

7. What type of probability distribution will most likely be used to analyze warranty repair needs on new cars in the following problem?

The service manager for a new automobile dealership reviewed dealership records of the past 20 sales of new cars to determine the number of warranty repairs he will be called on to perform in the next 30 days. Corporate reports indicate that the probability any one of their new cars needs a warranty repair in the first 30 days is 0.035. The manager assumes that calls for warranty repair are independent of one another and is interested in predicting the number of warranty repairs he will be called on to perform in the next 30 days for this batch of 20 new cars sold.

(a) Binomial distribution

(b) Poisson distribution

(c) Normal distribution

(d) None of the above

8. The quality control manager of Marilyn's Cookies is inspecting a batch of chocolate chip cookies. When the production process is in control, the mean number of chocolate chip parts per cookie is 6.0. The manager is interested in analyzing the probability that any particular cookie being inspected has fewer than 10.0 chip parts. What probability distribution should be used?

(a) Binomial distribution

(b) Poisson distribution

(c) Normal distribution

(d) None of the above

9. The smallest number of possible successes in a Poisson distribution is:

(a) 0

(b) 1

(c) n

(d) infinite

10. Based on past experience, the time you spend on emails per day has a mean of 30 minutes and a standard deviation of 10 minutes. To compute the probability of spending at least 12 minutes on emails, you can use what probability distribution?

(a) Binomial distribution

(b) Poisson distribution

(c) Normal distribution

(d) None of the above

11. A computer lab at a university has ten personal computers. Based on past experience, the probability that any one of them will require repair on a given day is 0.05. To find the probability that exactly two of the computers will require repair on a given day, you can use what probability distribution?

(a) Binomial distribution

(b) Poisson distribution

(c) Normal distribution

(d) None of the above

12. The mean number of customers who arrive per minute at any one of the checkout counters of a grocery store is 1.8. What probability distribution can be used to find out the probability that there will be no customers arriving at a checkout counter in the next minute?
    (a) Binomial distribution

    (b) Poisson distribution

    (c) Normal distribution

    (d) None of the above

13. A multiple-choice test has 25 questions. There are four choices for each question. A student who has not studied for the test decides to answer all questions by randomly choosing one of the four choices. What probability distribution can be used to compute his chance of correctly answering at least 15 questions?
    (a) Binomial distribution

    (b) Poisson distribution

    (c) Normal distribution

    (d) None of the above

14. Which of the following about the normal distribution are true?
    (a) Theoretically, the mean, median, and mode are the same.

    (b) About 99.7% of the values fall within three standard deviations from the mean.

    (c) It is defined by two characteristics μ and σ.

    (d) All of the above are true.

15. Which of the following about the normal distribution is not true?
    (a) Theoretically, the mean, median, and mode are the same.

    (b) About two-thirds of the observations fall within one standard deviation from the mean.

    (c) It is a discrete probability distribution.

    (d) Its parameters are the mean, μ, and standard deviation, σ.

16. The probability that Z is less than -1.0 is _____ the probability that Z is greater than $+1.0$.
    (a) less than

    (b) the same as

    (c) greater than

17. The normal distribution is _____ in shape:
    (a) right-skewed
    (b) left-skewed
    (c) symmetric

18. If a particular set of data is approximately normally distributed, you would find that approximately:
    (a) 2 of every 3 observations would fall between 1 standard deviation around the mean
    (b) 4 of every 5 observations would fall between 1.28 standard deviations around the mean
    (c) 19 of every 20 observations would fall between 2 standard deviations around the mean
    (d) All the above

19. Given that X is a normally distributed random variable with a mean of 50 and a standard deviation of 2, the probability that X is between 47 and 54 is _____.

Answer True or False:

20. Theoretically, the mean, median, and the mode are all equal for a normal distribution.

21. Another name for the mean of a probability distribution is its expected value.

22. The diameters of 100 randomly selected bolts follow a binomial distribution.

23. If the data values are normally distributed, the normal probability plot will follow a straight line.

Answers to Test Yourself Short Answers

1. d	13. a
2. c	14. d
3. a	15. c
4. d	16. b
5. a	17. c
6. a	18. d
7. a	19. 0.9104
8. b	20. True
9. a	21. True
10. c	22. False
11. a	23. True
12. b	

Problems

1. Given the following probability distributions:

Distribution A		Distribution B	
X	*P(X)*	*X*	*P(X)*
0	0.20	0	0.10
1	0.20	1	0.20
2	0.20	2	0.40
3	0.20	3	0.20
4	0.20	4	0.10

   (a) Compute the expected value of each distribution.
   (b) Compute the standard deviation of each distribution.
   (c) Compare the results of distributions *A* and *B*.

2. In the carnival game Under-or-Over-Seven, a pair of fair dice is rolled once, and the resulting sum determines whether the player wins or loses his or her bet. For example, the player can bet $1 that the sum will be under 7—that is, 2, 3, 4, 5, or 6. For this bet, the player wins $1 if the result is under 7 and loses $1 if the outcome equals or is greater than 7. Similarly, the player can bet $1 that the sum will be over 7— that is, 8, 9, 10, 11, or 12. Here, the player wins $1 if the result is over 7 but loses $1 if the result is 7 or under. A third method of play is to bet $1 on the outcome 7. For this bet, the player wins $4 if the result of the roll is 7 and loses $1 otherwise.
   (a) Construct the probability distribution representing the different outcomes that are possible for a $1 bet on being under 7.
   (b) Construct the probability distribution representing the different outcomes that are possible for a $1 bet on being over 7.
   (c) Construct the probability distribution representing the different outcomes that are possible for a $1 bet on 7.
   (d) Show that the expected long-run profit (or loss) to the player is the same, no matter which method of play is used.

3. The number of arrivals per minute at a bank located in the central business district of a large city was recorded over a period of 200 minutes with the following results:

Arrivals	Frequency
0	14
1	31
2	47
3	41
4	29
5	21
6	10
7	5
8	2

(a) Compute the expected number of arrivals per day.

(b) Compute the standard deviation.

4. Suppose that a judge's decisions are upheld by an appeals court 90% of the time. In her next ten decisions, what is the probability that
   (a) eight of her decisions are upheld by an appeals court?
   (b) all ten of her decisions are upheld by an appeals court?
   (c) eight or more of her decisions are upheld by an appeals court?

5. A venture capitalist firm that specializes in funding risky high-technology startup companies has determined that only one in ten of its companies is a "success" that makes a substantive profit within six years. Given this historical record, what is the probability that in the next three startups it finances:
   (a) The firm will have exactly one success?
   (b) Exactly two successes?
   (c) Less than two successes?
   (d) At least two successes?

6. A campus program enrolls undergraduate and graduate students. Of the students, 70% are undergraduates. If a random sample of four students is selected from the program to be interviewed about the introduction of a new fast-food outlet on the ground floor of the campus building, what is the probability that all four students selected are:
   (a) Undergraduate students?
   (b) Graduate students?

7. The number of power outages at a power plant has a Poisson distribution with a mean of four outages per year. What is the probability that in a year there will be
   (a) no power outages?
   (b) four power outages?
   (c) at least three power outages?

8. The quality control manager of Marilyn's Cookies is inspecting a batch of chocolate-chip cookies that has just been baked. If the production process is in control, the mean number of chip parts per cookie is 6.0. What is the probability that in any particular cookie being inspected there are
   (a) less than five chip parts?
   (b) exactly five chip parts?
   (c) five or more chip parts?
   (d) either four or five chip parts?

9. The U.S. Department of Transportation maintains statistics for mishandled bags per 1,000 airline passengers. In 2007, airlines had seven mishandled bags per 1,000 passengers (extracted from R. Yu, "Airline Performance nears Twenty Year Low," *USA Today*, April 8, 2008, p. B1). What is the probability that in the next 1,000 passengers, airlines will have
   (a) no mishandled bags?
   (b) at least one mishandled bag?
   (c) at least two mishandled bags?

10. Given that X is a normally distributed random variable with a mean of 50 and a standard deviation of 2, what is the probability that
   (a) X is between 47 and 54?
   (b) X is less than 55?
   (c) There is a 90% chance that X will be less than what value?

11. A set of final examination grades in an introductory statistics course is normally distributed, with a mean of 73 and a standard deviation of 8.
   (a) What is the probability of getting a grade below 91 on this exam?
   (b) What is the probability that a student scored between 65 and 89?
   (c) The probability is 5% that a student taking the test scores higher than what grade?
   (d) If the professor grades on a curve (that is, gives A's to the top 10% of the class, regardless of the score), are you better off with a grade of 81 on this exam or a grade of 68 on a different exam, where the mean is 62 and the standard deviation is 3? Explain.

12. The owner of a fish market determined that the mean weight for salmon is 12.3 pounds with a standard deviation of 2 pounds.

Assuming the weights of salmon are normally distributed, what is the probability that

(a) a randomly selected salmon will weigh between 12 and 15 pounds?

(b) a randomly selected salmon will weigh less than 10 pounds?

(c) 95% of the salmon will weigh between what two values?

RestCost

13. On page 106, a normal probability plot was constructed for the cost of a meal at city restaurants. Construct a normal probability plot of the cost of a meal at suburban restaurants. Do you think that the cost of a meal at suburban restaurants is normally distributed? Explain.

NBACost

14. Is the cost of attending an NBA basketball game normally distributed? Construct a normal probability plot of the cost of attending an NBA basketball game. Do you think that the cost of attending an NBA basketball game is normally distributed? Explain.

Chemical

15. Is the viscosity of a certain chemical normally distributed? Construct a normal probability plot of the viscosity of the chemical. Do you think that the viscosity of the chemical is normally distributed? Explain.

Answers to Test Yourself Problems

1. (a) A: $\mu = 2$, B: $\mu = 2$.

   (b) A: $\sigma = 1.414$, B: $\sigma = 1.095$.

   (c) Distribution A is uniform and symmetric; Distribution B is symmetric and has a smaller standard deviation than distribution A.

2. (a)

X	$P(X)$
$\$ - 1$	21/36
$\$ + 1$	15/36

   (b)

X	$P(X)$
$\$ - 1$	21/36
$\$ + 1$	15/36

   (c)

X	$P(X)$
$\$ - 1$	30/36
$\$ + 4$	6/36

   (d) $\$-0.167$ for each method of play.

3. (a) 2.90

   (b) 1.772

4. (a) 0.1937

   (b) 0.3487

   (c) 0.9298

5. (a) 0.243

   (b) 0.027

   (c) 0.972

   (d) 0.028

6. (a) 0.2401

   (b) 0.0081

7. (a) 0.0183

   (b) 0.1954

   (c) 0.7619

8. (a) 0.2851

   (b) 0.1606

   (c) 0.7149

   (d) 0.2945

9. (a) 0.0009

   (b) 0.9991

   (c) 0.9927

10. (a) 0.9104

   (b) 0.9938

   (c) 52.5631

11. (a) 0.9878

   (b) 0.8185

   (c) 86.16%

   (d) Option 1: Because your score of 81% on this exam represents a Z score of 1.00, which is below the minimum Z score of 1.28, you will not earn an A grade on the exam under this grading option. Option 2: Because your score of 68% on this exam represents a Z score of 2.00, which is well above the minimum Z score of 1.28, you will earn an A grade on the exam under this grading option. You should prefer option 2.

12. (a) 0.4711

   (b) 0.1251

   (c) 8.38 and 16.22

13. The cost of a meal at suburban restaurants is approximately normally distributed because the normal probably plot is approximately a straight line.

14. The cost of attending an NBA basketball game appears to be right-skewed.

15. The viscosity of a chemical is approximately normally distributed because the normal probability plot is approximately a straight line.

References

1. Berenson, M. L., D. M. Levine, and T. C. Krehbiel. *Basic Business Statistics: Concepts and Applications*, Eleventh Edition. Upper Saddle River, NJ: Prentice Hall, 2009.

2. Levine, D. M., T. C. Krehbiel, and M. L. Berenson. *Business Statistics: A First Course*, Fifth Edition. Upper Saddle River, NJ: Prentice Hall, 2010.

3. Levine, D. M., D. Stephan, T. C. Krehbiel, and M. L. Berenson. *Statistics for Managers Using Microsoft Excel*, Fifth Edition. Upper Saddle River, NJ: Prentice Hall, 2008.

4. Levine, D. M., P. P. Ramsey, and R. K. Smidt. *Applied Statistics for Engineers and Scientists Using Microsoft Excel and Minitab*. Upper Saddle River, NJ: Prentice Hall, 2001.

5. Microsoft Excel 2007. Redmond, WA: Microsoft Corporation, 2006.

CHAPTER

6

Sampling Distributions and Confidence Intervals

6.1   Sampling Distributions
6.2   Sampling Error and Confidence Intervals
6.3   Confidence Interval Estimate for the Mean Using
        the *t* Distribution (σ Unknown)
6.4   Confidence Interval Estimation for Categorical
        Variables
        Important Equations
        One-Minute Summary
        Test Yourself

In Section 1.3, you learned that you can use **inferential statistics** to make conclusions about a large set of data called the **population**, based on a subset of the data, called the **sample**. Specifically, inferential statistical methods use a single sample to calculate a sample statistic from which conclusions about a population parameter are made. Using a small part to make conclusions about a whole larger thing sometimes strikes people as being counterintuitive, but the ability to make such conclusions using inferential methods is one of the important applications of statistics.

As a first step towards understanding inferential statistics, this chapter introduces the concepts of a sampling distribution and a confidence interval. You also learn how to construct confidence interval estimates for the mean and the proportion. By the end of the chapter, you will begin to see how a small part of the whole can enable you to make plausible inferences about the whole.

6.1 Sampling Distributions

Your knowledge of **sampling distributions**, when combined with the probability and probability distribution concepts of the previous two chapters, provides you with the theoretical justifications that enable you to arrive at conclusions about an entire population based only on a single sample.

Sampling Distribution

CONCEPT   The distribution of a sample statistic, such as the mean, for all possible samples of a given size n.

EXAMPLES   Sampling distribution of the mean, sampling distribution of the proportion.

INTERPRETATION   Consider a population that includes 1,000 items. The sampling distribution of the mean for samples of 15 items consists of the mean of every single different sample of 15 items from the population. Imagine the distribution of all the means that could possibly occur: Some means would be smaller than others, some would be larger than others, and many would have similar values.

Calculating the means for all the samples would be an involved and time-consuming task. Fortunately, you do not have to develop specific sampling distributions yourself because statisticians have extensively studied sampling distributions for many different statistics, including the widely used sampling distribution for the mean and the sampling distribution for the proportion. These well-known sampling distributions are used extensively, starting in this chapter and continuing through Chapter 9, as a basis for inferential statistics.

important point

As you read through the remainder of this chapter and Chapters 7 through 9, remember these facts associated with sampling distributions:

- Every sample statistic has a sampling distribution.
- A specific sample statistic is used to estimate its corresponding population characteristic.
- Each sample statistic of interest is associated with a specific sampling distribution.

Sampling Distribution of the Mean and the Central Limit Theorem

The mean is the most widely used measure in statistics. Recall from Section 3.1 that the mean is equal to the sum of the data values, divided by the number of data values that were summed, and that because the mean uses all the data values, it has one great weakness: An individual extreme value can distort the mean.

Through several insights including the observation that the probability of getting such a distorted mean is relatively low, whereas the probability of getting a mean similar to many other sample means is much greater, statisticians have developed the **central limit theorem**. This theorem states that

important point

> Regardless of the shape of the distribution of the individual values in the population, as the sample size *gets large enough*, the sampling distribution of the mean can be approximated by a normal distribution.

As a general rule, statisticians have found that for many population distributions, a sample size of at least 30 is "large enough." However, you can apply the central limit theorem for smaller sample sizes if the distribution is known to be approximately bell shaped. In the uncommon case in which the distribution is extremely skewed or has more than one mode, sample sizes larger than 30 might be needed in order to apply the theorem. Figure 6.1 contains the sampling distribution of the mean for three different populations. For each population, the sampling distribution of the sample mean is shown for all samples of $n = 2$, $n = 5$, and $n = 30$.

Panel A illustrates the sampling distribution of the mean selected from a population that is normally distributed. When the population is normally distributed, the sampling distribution of the mean is normally distributed regardless of the sample size. If the sample size increases, the variability of the sample mean from sample to sample decreases.

Panel B displays the sampling distribution from a population with a uniform (or rectangular) distribution. When samples of size $n = 2$ are selected, a **central limiting** effect is already working in which more sample means are in the center than there are individual values. For $n = 5$, the sampling distribution is bell shaped and approximately normal. When $n = 30$, the sampling distribution appears to be very similar to a normal distribution. In general, the larger the sample size, the more closely the sampling distribution follows a normal distribution. As with all cases, the mean of each sampling distribution is equal to the mean of the population, and the variability decreases as the sample size increases.

Panel C presents an exponential distribution. This population is heavily skewed to the right. When $n = 2$, the sampling distribution is still highly skewed to the right, but less so than the distribution of the population. For $n = 5$, the sampling distribution is approximately symmetric with only a slight skew to the right. When $n = 30$, the sampling distribution appears to be approximately normally distributed. Again, the mean of each sampling distribution is equal to the mean of the population and the variability decreases as the sample size increases.

important point

From this figure you are able to reach the following conclusions about the sampling distribution of the mean:

- For most population distributions, regardless of shape, the sampling distribution of the mean is approximately normally distributed if samples of at least 30 observations are selected.

- If the population distribution is fairly symmetrical, the sampling distribution of the mean is approximately normally distributed if samples of at least 15 observations are selected.
- If the population is normally distributed, the sampling distribution of the mean is normally distributed regardless of the sample size.

Figure 6.1

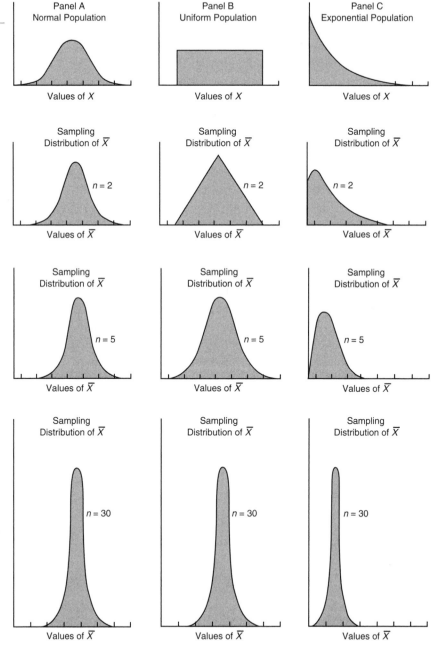

Sampling Distribution of the Proportion

Recall from Section 5.2 that you can use a binomial distribution to determine probabilities for categorical variables that have only two categories, traditionally labeled "success" and "failure." As the sample size increases for such variables, you can use the normal distribution to approximate the sampling distribution of the number of successes or the proportion of successes. Specifically, as a general rule, you can use the normal distribution to approximate the binomial distribution when the number of successes and the number of failures are *each* at least five. For most cases in which you are estimating the proportion, the sample size is more than sufficient to meet the conditions for using the normal approximation.

6.2 Sampling Error and Confidence Intervals

Taking one sample and computing the results of a sample statistic, such as the mean, creates a **point estimate** of the population parameter. This single estimate will almost certainly be different if another sample is selected.

OrderTimePopulation

For example, consider the following table that records the results of taking 20 samples of n = 15 selected from a population of N = 200 order-filling times. This population has a population mean = 69.637 and a population standard deviation = 10.411.

SamplesofOrderTimes

Sample	Mean	Standard Deviation	Minimum	Median	Maximum	Range
1	66.12	9.21	47.20	65.00	87.00	39.80
2	73.30	12.48	52.40	71.10	101.10	48.70
3	68.67	10.78	54.00	69.10	85.40	31.40
4	69.95	10.57	54.50	68.00	87.80	33.30
5	73.27	13.56	54.40	71.80	101.10	46.70
6	69.27	10.04	50.10	70.30	85.70	35.60
7	66.75	9.38	52.40	67.30	82.60	30.20
8	68.72	7.62	54.50	68.80	81.50	27.00
9	72.42	9.97	50.10	71.90	88.90	38.80
10	69.25	10.68	51.10	66.50	85.40	34.30
11	72.56	10.60	60.20	69.10	101.10	40.90
12	69.48	11.67	49.10	69.40	97.70	48.60

Sample	Mean	Standard Deviation	Minimum	Median	Maximum	Range
13	64.65	9.71	47.10	64.10	78.50	31.40
14	68.85	14.42	46.80	69.40	88.10	41.30
15	67.91	8.34	52.40	69.40	79.60	27.20
16	66.22	10.18	51.00	66.40	85.40	34.40
17	68.17	8.18	54.20	66.50	86.10	31.90
18	68.73	8.50	57.70	66.10	84.40	26.70
19	68.57	11.08	47.10	70.40	82.60	35.50
20	75.80	12.49	56.70	77.10	101.10	44.40

From these results, you can observe the following:

- The sample statistics differ from sample to sample. The sample means vary from 64.65 to 75.80, the sample standard deviations vary from 7.62 to 14.42, the sample medians vary from 64.10 to 77.10, and the sample ranges vary from 26.70 to 48.70.

- Some of the sample means are higher than the population mean of 69.637, and some of the sample means are lower than the population mean.

- Some of the sample standard deviations are higher than the population standard deviation of 10.411, and some of the sample standard deviations are lower than the population standard deviation.

- The variation in the sample range from sample to sample is much greater than the variation in the sample standard deviation.

You should realize that sample statistics almost always vary from sample to sample. This expected variation is called the **sampling error**.

Sampling Error

CONCEPT   The variation that occurs due to selecting a single sample from the population.

EXAMPLE   In polls, the plus-or-minus margin of the results; as in "42%, plus or minus 3%, said they were likely to vote for the incumbent."

INTERPRETATION   The size of the sampling error is primarily based on the variation in the population itself and on the size of the sample selected. Larger samples will have less sampling error, but will be more costly to take.

In practice, only one sample is used as the basis for estimating a population parameter. To account for the differences in the results from sample to sample, statisticians have developed the concept of a **confidence interval estimate**, which indicates the likelihood that a stated interval with a lower and upper limit properly estimates the parameter.

Confidence Interval Estimate

CONCEPT   An estimate of a population parameter stated as a range with a lower and upper limit with a specific degree of certainty.

INTERPRETATION   All that you need to know to develop a confidence interval estimate is the sample statistic used to estimate the population parameter and the sampling distribution for the sample statistic. This is always true regardless of the population parameter being estimated.

Because you are estimating an interval using one sample and not precisely determining a value, there is no way that you can be 100% certain that your interval correctly estimates the population parameter as noted earlier and illustrated by WORKED-OUT PROBLEM 1. However, by setting the level of certainty to a value below 100%, you can use the interval estimate to make plausible inferences about the population with a given degree of certainty.

There is a trade-off between the level of confidence and the width of the interval. For a given sample size, if you want more confidence that your interval will be correct, you will have a wider interval and therefore a less precise estimate.

Given this trade-off, what level of certainty should you use? As expressed as a percentage, the most common percentage used is 95%. If more confidence is needed, 99% is typically used; if less confidence is needed, 90% is typically used.

Because of this factor, the degree of certainty, or **confidence**, must always be stated when reporting an interval estimate. When you hear an "interval estimate with 95% confidence," or simply, a "95% confidence interval estimate," you can conclude that if all possible samples of the same size n were selected, 95% of them would include the population parameter somewhere within the interval and 5% would not.

WORKED-OUT PROBLEM 1   You want to develop 95% confidence interval estimates for the mean from 20 samples of size 15 for the order-filling data presented on page [xref]. Unlike most real-life problems, the population mean, $\mu = 69.637$, and the population standard deviation, $\sigma = 10.411$, are already known, so the confidence interval estimate for the mean developed from each sample can be compared to the actual value of the population mean.

95% Confidence Interval Estimates from 20 Samples of n = 15 Selected from a Population of N = 200 with $\mu = 69.637$ *and* $\sigma = 10.411$

Sample	Mean	Standard Deviation	Lower Limit	Upper Limit
1	66.12	9.21	60.85	71.39
2	73.30	12.48	68.03	78.57
3	68.67	10.78	63.40	73.94
4	69.95	10.57	64.68	75.22
5	73.27	13.56	68.00	78.54
6	69.27	10.04	64.00	74.54
7	66.75	9.38	61.48	72.02
8	68.72	7.62	63.45	73.99
9	72.42	9.97	67.15	77.69
10	69.25	10.68	63.98	74.52
11	72.56	10.60	67.29	77.83
12	69.48	11.67	64.21	74.75
13	64.65	9.71	59.38	69.92
14	68.85	14.42	63.58	74.12
15	67.91	8.34	62.64	73.18
16	66.22	10.18	60.95	71.49
17	68.17	8.18	62.90	73.44
18	68.73	8.50	63.46	74.00
19	68.57	11.08	63.30	73.84
20	75.80	12.49	70.53	81.07

From the results, you can conclude the following:

- For sample 1, the sample mean is 66.12, the sample standard deviation is 9.21, and the interval estimate for the population mean is 60.85 to 71.39. This enables you to conclude with 95% certainty that the population mean is between 60.85 and 71.39. This is a correct estimate because the population mean of 69.637 is included within this interval.

- Although their sample means and standard deviations differ, the confidence interval estimates for samples 2 through 19 lead to interval estimates that include the population mean value.

- For sample 20, the sample mean is 75.80, the sample standard deviation is 12.49, and the interval estimate for the population mean is 70.53 to 81.07 (highlighted in the results). This is an incorrect estimate because the population mean of 69.637 is not included within this interval.

You might realize that these results are not surprising because the percentage of correct results (19 out of 20) is 95%, just as statistical theory would claim. Of course, with other specific sets of 20 samples, the percentage of correct results might not be exactly 95%—it could be higher or lower—but in the long run, 95% of all samples do result in a correct estimate.

6.3 Confidence Interval Estimate for the Mean Using the *t* Distribution (σ Unknown)

The most common confidence interval estimate involves estimating the mean of a population. In virtually all cases, the population mean is estimated from sample data in which only the sample mean and sample standard deviation—and not the population standard deviation—are known. To overcome this complication, statisticians (see Reference 1) have developed the *t* distribution.

t Distribution

CONCEPT The sampling distribution that allows you to develop a confidence interval estimate of the mean using the sample standard deviation.

important point ✏️

INTERPRETATION The *t* distribution assumes that the variable being studied is normally distributed. In practice, however, as long as the sample size is large enough and the population is not very skewed, the *t* distribution can be used to estimate the population mean when the population standard deviation σ is unknown. You should be concerned about the validity of the confidence interval primarily when dealing with a small sample size and a skewed population distribution. The assumption of normality in the population can be assessed by evaluating the shape of the sample data using a histogram, box-and-whisker plot, or normal probability plot.

WORKED-OUT PROBLEM 2 You want to undertake a study that compares the cost for a restaurant meal in a major city to the cost of a similar meal in the suburbs outside the city. You collect data about the cost of a meal per person from a sample of 50 city restaurants and 50 suburban restaurants as follows:

RestCost

City Cost Data

13	21	22	22	24	25	26	26	26	26
30	32	33	34	34	35	35	35	35	36
37	37	39	39	39	40	41	41	41	42
43	44	45	46	50	50	51	51	53	53
53	55	57	61	62	62	62	66	68	75

Suburban Cost Data

21	22	25	25	26	26	27	27	28	28
28	29	31	32	32	35	35	36	37	37
37	38	38	38	39	40	40	41	41	41
42	42	43	44	47	47	47	48	50	50
50	50	50	51	52	53	58	62	65	67

Spreadsheet results for the confidence interval estimate of the population mean for the mean cost of a meal in city restaurants and in suburban restaurants are as follows:

	A	B	C
1	Confidence Interval Estimate for the Mean		
2			
3	Data	City	Suburban
4	Sample Standard Deviation	13.8891	11.1355
5	Sample Mean	41.46	39.96
6	Sample Size	50	50
7	Confidence Level	95%	95%
8			
9	Intermediate Calculations		
10	Standard Error of the Mean	1.9642	1.5748
11	Degrees of Freedom	49	49
12	t Value	2.0096	2.0096
13	Interval Half Width	3.9472	3.1647
14			
15	Confidence Interval	City	Suburban
16	Interval Lower Limit	37.51	36.80
17	Interval Upper Limit	45.41	43.12

To evaluate the assumption of normality necessary to use these estimates, you create box-and-whisker plots for the cost of the restaurant meals in the city and in the suburban areas (these were shown in Chapter 3 "Descriptive Statistics," on page 61).

You can see that the box-and-whisker plots have some right-skewness because the tail on the right is longer than the one on the left. However, given the relatively large sample size, you can conclude that any departure from the normality assumption will not seriously affect the validity of the confidence interval estimate.

Based on these results, with 95% confidence you can conclude that the mean cost of 9 meal is between $37.51 and $45.41 for city restaurants and between $36.80 and $43.12 for suburban restaurants

equation blackboard (optional)

interested in math?

You use the symbols $\overline{X}$ (sample mean), μ (population mean), S (sample standard deviation), and n (sample size), all introduced earlier, and include the new symbol t_{n-1}, which represents the critical value of the t distribution with $n-1$ degrees of freedom for an area of $\alpha/2$ in the upper tail, to express the confidence interval for the amean in cases in which the population standard deviation, σ, is unknown as a formula. For the symbol t_{n-1}:

$n-1$ is one less than the sample size and α is equivalent to 1 minus the confidence percentage. For 95% confidence, α is 0.05 (1–0.95), so the upper tail area is 0.025.

Using these symbols creates the following equation:

$$\overline{X} \pm t_{n-1} \frac{S}{\sqrt{n}}$$

or expressed as a range

$$\overline{X} - t_{n-1} \frac{S}{\sqrt{n}} \leq \mu \leq \overline{X} + t_{n-1} \frac{S}{\sqrt{n}}$$

For the city restaurant meal cost example of this section, $\overline{X}$ = 41.46, and S = 13.8891, and because the sample size is 50, there are 49 degrees of freedom. Given 95% confidence, α is 0.05, and the area in the upper tail of the t distribution is 0.025 (0.05/2). Using Table C.2 (in Appendix C), the critical value for the row with 49 degrees of freedom and the column with an area of 0.025 is 2.0096. Substituting these values yields the following result:

$$\overline{X} \pm t_{n-1} \frac{S}{\sqrt{n}}$$

$$= 41.46 \pm (2.0096) \frac{13.8891}{\sqrt{50}}$$

$$= 41.46 \pm 3.9472$$

$$37.51 \leq \mu \leq 45.41$$

The interval is estimated to be between $37.51 and $45.41 with 95% confidence.

calculator keys

Confidence Interval Estimate for the Mean When σ Is Unknown

Press [STAT][◀] to display the Tests menu and select 8:TInterval and press [Enter] to display the TInterval screen. In this screen (see the following left figure), select **Stats** as the **Inpt** type and press [ENTER].

Enter values for the sample mean $\bar{X}$, the sample standard deviation S_x, and the sample size, *n*. Then enter the confidence level (**C-Level**) as the decimal fraction equivalent to a percentage. (Enter .95 for 95% confidence.) Select **Calculate** and press [ENTER] to display the results screen. In this screen (see the following right figure), the lower and upper limits of confidence interval estimate appear as a pair of values enclosed in parentheses (37.513, 45.407, in the figure).

```
TInterval          TInterval
 Inpt:Data Stats   (37.513,45.407)
 x:41.46           x=41.46
 Sx:13.8891        Sx=13.8891
 n:50              n=50
 C-Level:.95
 Calculate
```

spreadsheet solution

Confidence Interval Estimate for the Mean When σ Is Unknown

Chapter 6 Sigma Unknown determines the confidence interval estimate for the mean when σ is unknown based on the values for the sample standard deviation, the sample mean, the sample size, and the confidence level (as a percentage) that you enter.

Spreadsheet Tip FT6 in Appendix D explains the spreadsheet function that calculates the critical value of the *t* distribution (see the earlier EQUATION BLACKBOARD) that is used in determining the confidence interval estimate.

6.4 Confidence Interval Estimation for Categorical Variables

For a categorical variable, you can develop a confidence interval to estimate the proportion of successes in a given category.

Confidence Interval Estimation for the Proportion

CONCEPT   The sampling distribution of the proportion that allows you to develop a confidence interval estimate of the proportion using the sample proportion of successes, p. The sample statistic p follows a binomial distribution that can be approximated by the normal distribution for most studies.

EXAMPLE   The proportion of voters who would vote for a certain candidate in an election, the proportion of consumers who prefer a particular brand of soft drink, the proportion of medical tests in a hospital that need to be repeated.

INTERPRETATION   This type of confidence interval estimate uses the sample proportion of successes, p, equal to the number of successes divided by the sample size, to estimate the population proportion. (Categorical variables have no population means.)

For a given sample size, confidence intervals for proportions are wider than those for numerical variables. With continuous variables, the measurement on each respondent contributes more information than for a categorical variable. In other words, a categorical variable with only two possible values is a very crude measure compared with a continuous variable, so each observation contributes only a little information about the parameter being estimated.

WORKED-OUT PROBLEM 3   You want to estimate the proportion of people who take work with them on vacation. In a recent survey by **CareerJournal.com** (data extracted from P. Kitchen, "Can't Turn It Off," *Newsday*, October 20, 2006, pp. F4–F5), 158 of 473 employees responded that they typically took work with them on vacation.

Based on the 95% confidence interval estimate prepared in Microsoft Excel for the proportion of people who take work with them on vacation (see the following figure), you estimate that between 29.15% and 37.65% of people take work with them on vacation.

	A	B
1	**Confidence Interval Estimate for the Proportion**	
2		
3	**Data**	
4	**Sample Size**	473
5	**Number of Successes**	158
6	**Confidence Level**	95%
7		
8	Intermediate Calculations	
9	Sample Proportion	0.3340
10	Z Value	-1.9600
11	Standard Error of the Proportion	0.0217
12	Interval Half Width	0.0425
13		
14	**Confidence Interval**	
15	**Interval Lower Limit**	0.2915
16	**Interval Upper Limit**	0.3765

equation
blackboard
(optional)

interested
in
math?

You use the symbols p (sample proportion of success), n (sample size), and Z (Z score), previously introduced, and the symbol π for the population proportion, to assemble the equation for the confidence interval estimate for the proportion:

$$p \pm Z\sqrt{\frac{p(1-p)}{n}}$$

or expressed as a range

$$p - Z\sqrt{\frac{p(1-p)}{n}} \leq \pi \leq p + Z\sqrt{\frac{p(1-p)}{n}}$$

Z = Critical value from the normal distribution

For WORKED-OUT PROBLEM 3, $n = 473$ and $p = 158/473 = 0.334$. For a 95% level of confidence, the lower tail area of 0.025 provides a Z value from the normal distribution of –1.96, and the upper tail area of 0.025 provides a Z value from the normal distribution of +1.96. Substituting these numbers into the preceding equation yields the following result:

$$p \pm Z\sqrt{\frac{p(1-p)}{n}}$$

$$= 0.334 \pm (1.96)\sqrt{\frac{(0.334)(0.666)}{473}}$$

$$= 0.334 \pm (1.96)((0.022))$$

$$= 0.334 \pm 0.043$$

$$0.291 \leq \pi \leq 0.377$$

The proportion of nonconforming newspapers is estimated to be between 29.1% and 37.7%.

calculator keys

Confidence Interval Estimate for the Proportion

Press [STAT][◄] to display the Tests menu and select
A:1-PropZInt to display the 1-PropZInt screen. In this screen,
enter values for the number of successes *x*, the sample size *n*,
and the confidence level (**C-Level**) as the decimal fraction
equivalent to a percentage (.95 for 95%). Select **Calculate** and
press [ENTER] to display the results screen. The lower and
upper limits of the confidence interval estimate will appear as
a pair of values enclosed in parentheses.

```
1-PropZInt
x:158
n:473
C-Level:.95
Calculate
```

```
1-PropZInt
(.29153,.37654)
p=.334038055
n=473
```

spreadsheet solution

Confidence Interval Estimate for the Proportion

Chapter 6 Proportion determines the confidence interval estimate for the proportion based on the values for the sample
size, the number of successes, and the confidence level (as a
percentage) that you enter. Spreadsheet Tip FT7 in Appendix
D explains the spreadsheet function that calculates the critical
value of the normal distribution that is used in determining
the confidence interval estimate.

Important Equations

Confidence interval for the mean with **σ** unknown:

$$\overline{X} \pm t_{n-1} \frac{S}{\sqrt{n}}$$

(6.1)   *or*

$$\overline{X} - t_{n-1} \frac{S}{\sqrt{n}} \le \mu \le \overline{X} + t_{n-1} \frac{S}{\sqrt{n}}$$

Confidence interval estimate for the proportion:

$$p \pm Z \sqrt{\frac{p(1-p)}{n}}$$

(6.2)   *or*

$$p - Z \sqrt{\frac{p(1-p)}{n}} \le \pi \le p + Z \sqrt{\frac{p(1-p)}{n}}$$

One-Minute Summary

For what type of variable are you developing a confidence interval estimate?

- If a numerical variable, use the confidence interval estimate for the mean.

- If a categorical variable, use the confidence interval estimate for the proportion.

Test Yourself
Short Answers

1. The sampling distribution of the mean can be approximated by the normal distribution:
    (a) as the number of samples gets "large enough"
    (b) as the sample size (number of observations in each sample) gets large enough
    (c) as the size of the population standard deviation increases
    (d) as the size of the sample standard deviation decreases

2. The sampling distribution of the mean requires _____ sample size to reach a normal distribution if the population is skewed than if the population is symmetrical.
   (a) the same
   (b) a smaller
   (c) a larger
   (d) The two distributions cannot be compared.

3. Which of the following is true regarding the sampling distribution of the mean for a large sample size?
   (a) It has the same shape and mean as the population.
   (b) It has a normal distribution with the same mean as the population.
   (c) It has a normal distribution with a different mean from the population.

4. For samples of $n = 30$, for most populations, the sampling distribution of the mean will be approximately normally distributed:
   (a) regardless of the shape of the population
   (b) if the shape of the population is symmetrical
   (c) if the standard deviation of the mean is known
   (d) if the population is normally distributed

5. For samples of $n = 1$, the sampling distribution of the mean will be normally distributed:
   (a) regardless of the shape of the population
   (b) if the shape of the population is symmetrical
   (c) if the standard deviation of the mean is known
   (d) if the population is normally distributed

6. A 99% confidence interval estimate can be interpreted to mean that:
   (a) If all possible samples are taken and confidence interval estimates are developed, 99% of them would include the true population mean somewhere within their interval.
   (b) You have 99% confidence that you have selected a sample whose interval does include the population mean.
   (c) Both a and b are true.
   (d) Neither a nor b is true.

7. Which of the following statements is false?
   (a) There is a different critical value for each level of alpha (α).
   (b) Alpha (α) is the proportion in the tails of the distribution that is outside the confidence interval.
   (c) You can construct a 100% confidence interval estimate of μ.
   (d) In practice, the population mean is the unknown quantity that is to be estimated.

8. Sampling distributions describe the distribution of:
    (a) parameters
    (b) statistics
    (c) both parameters and statistics
    (d) neither parameters nor statistics

9. In the construction of confidence intervals, if all other quantities are unchanged, an increase in the sample size will lead to a _____ interval.
    (a) narrower
    (b) wider
    (c) less significant
    (d) the same

10. As an aid to the establishment of personnel requirements, the manager of a bank wants to estimate the mean number of people who arrive at the bank during the two-hour lunch period from 12 noon to 2 p.m. The director randomly selects 64 different two-hour lunch periods from 12 noon to 2 p. m. and determines the number of people who arrive for each. For this sample, $\bar{X} = 49.8$ and $S = 5$. Which of the following assumptions is necessary in order for a confidence interval to be valid?
    (a) The population sampled from has an approximate normal distribution.
    (b) The population sampled from has an approximate t distribution.
    (c) The mean of the sample equals the mean of the population.
    (d) None of these assumptions are necessary.

11. A university dean is interested in determining the proportion of students who are planning to attend graduate school. Rather than examine the records for all students, the dean randomly selects 200 students and finds that 118 of them are planning to attend graduate school. The 95% confidence interval for p is 0.59 ±0.07. Interpret this interval.
    (a) You are 95% confident that the true proportion of all students planning to attend graduate school is between 0.52 and 0.66.
    (b) There is a 95% chance of selecting a sample that finds that between 52% and 66% of the students are planning to attend graduate school.
    (c) You are 95% confident that between 52% and 66% of the sampled students are planning to attend graduate school.
    (d) You are 95% confident that 59% of the students are planning to attend graduate school.

12. In estimating the population mean with the population standard deviation unknown, if the sample size is 12, there will be _____ degrees of freedom.

13. The Central Limit Theorem is important in statistics because
   (a) It states that the population will always be approximately normally distributed.
   (b) It states that the sampling distribution of the sample mean is approximately normally distributed for a large sample size n regardless of the shape of the population.
   (c) It states that the sampling distribution of the sample mean is approximately normally distributed for any population regardless of the sample size.
   (d) For any sized sample, it says the sampling distribution of the sample mean is approximately normal.

14. For samples of $n = 15$, the sampling distribution of the mean will be normally distributed:
   (a) regardless of the shape of the population
   (b) if the shape of the population is symmetrical
   (c) if the standard deviation of the mean is known
   (d) if the population is normally distributed

Answer True or False:

15. Other things being equal, as the confidence level for a confidence interval increases, the width of the interval increases.

16. As the sample size increases, the effect of an extreme value on the sample mean becomes smaller.

17. A sampling distribution is defined as the probability distribution of possible sample sizes that can be observed from a given population.

18. The t distribution is used to construct confidence intervals for the population mean when the population standard deviation is unknown.

19. In the construction of confidence intervals, if all other quantities are unchanged, an increase in the sample size will lead to a wider interval.

20. The confidence interval estimate that is constructed will always correctly estimate the population parameter.

Answers to Test Yourself Short Answers

1. b	8. b	15. True
2. c	9. a	16. True
3. b	10. d	17. False
4. a	11. a	18. True
5. d	12. 11	19. False
6. c	13. b	20. False
7. c	14. b	

Problems

MoviePrices

1. The data in the file **MoviePrices** contain the price for two tickets with online service charges, large popcorn, and two medium soft drinks at a sample of six Ftheatre chains:

$36.15 $31.00 $35.05 $40.25 $33.75 $43.00

Source: Extracted from K. Kelly, "The Multiplex Under Siege," *The Wall Street Journal*, December 24–25, 2005, pp. P1, P5.

Construct a 95% confidence interval estimate of the population mean price for two tickets with online service charges, large popcorn, and two medium soft drinks.

Sushi

2. Tuna sushi was purchased from 13 Manhattan restaurants and tested for mercury. The number of pieces it would take to reach what the Environmental Protection Agency considers to be an acceptable level to be regularly consumed was as follows:

8.6 2.6 1.6 5.2 7.7 4.7 6.4 6.2 3.6 4.9 9.9 3.3 4.1

Source: Data extracted from M. Burros, "High levels of Mercury Found in Tuna Sushi Sold in Manhattan," *The New York Times*, January 23, 2008, p. A1, A23.

Construct a 95% confidence interval estimate of the population mean number of pieces it would take to reach what the Environmental Protection Agency considers to be an acceptable level to be regularly consumed.

Chemical

3. The following data represent the viscosity (friction, as in automobile oil) taken from 120 manufacturing batches (ordered from lowest viscosity to highest viscosity).

12.6	12.8	13.0	13.1	13.3	13.3	13.4	13.5	13.6	13.7
13.7	13.7	13.8	13.8	13.9	13.9	14.0	14.0	14.0	14.1
14.1	14.1	14.2	14.2	14.2	14.3	14.3	14.3	14.3	14.3
14.3	14.4	14.4	14.4	14.4	14.4	14.4	14.4	14.4	14.5
14.5	14.5	14.5	14.5	14.5	14.6	14.6	14.6	14.7	14.7
14.8	14.8	14.8	14.8	14.9	14.9	14.9	14.9	14.9	14.9
14.9	15.0	15.0	15.0	15.0	15.1	15.1	15.1	15.1	15.2
15.2	15.2	15.2	15.2	15.2	15.2	15.2	15.3	15.3	15.3
15.3	15.3	15.4	15.4	15.4	15.4	15.5	15.5	15.6	15.6
15.6	15.6	15.6	15.7	15.7	15.7	15.8	15.8	15.9	15.9
16.0	16.0	16.0	16.0	16.1	16.1	16.1	16.2	16.3	16.4
16.4	16.5	16.5	16.6	16.8	16.9	16.9	17.0	17.6	18.6

(a) Construct a 95% confidence interval estimate of the population mean viscosity.

(b) Do you need to assume that the population viscosity is normally distributed to construct the confidence interval estimate for the mean viscosity? Explain.

4. In a survey of 1,000 airline travelers, 760 responded that the airline fee that is most unreasonable was additional charges to redeem points/miles (extracted from "Snapshots: Which Airline Fee Is Most Unreasonable?", *USA Today*, December 2, 2008, p. B1). Construct a 95% confidence interval estimate of the population proportion of airline travelers who think that the airline fee that is most unreasonable was additional charges to redeem points/miles.

5. In a survey of 2,395 adults, 1,916 reported that emails are easy to misinterpret, but only 1,269 reported that telephone conversations are easy to misinterpret (extracted from "Snapshots: Open to Misinterpretation," *USA Today*, July 17, 2007, p. 1D).
   (a) Construct a 95% confidence interval estimate of the population proportion of adults who report that emails are easy to misinterpret.
   (b) Construct a 95% confidence interval estimate of the population proportion of adults who report that telephone conversations are easy to misinterpret.
   (c) Compare the results of (a) and (b).

6. You want to estimate the proportion of newspapers printed that have a nonconforming attribute, such as excessive ruboff, improper page setup, missing pages, or duplicate pages. A random sample of $n = 200$ newspapers is selected from all the newspapers printed during a single day. In this sample, 35 contain some type of nonconformance. Construct a 90% confidence interval for the proportion of newspapers printed during the day that have a nonconforming attribute.

Answers to Test Yourself Problems

1. $\$31.93 \leq \mu \leq \$41.14$

2. $3.84 \leq \mu \leq 6.75$

3. (a) $14.80 \leq \mu \leq 15.16$

   (b) Because the sample size is large, at $n = 120$, the use of the t distribution is appropriate.

4. $0.7335 \leq \pi \leq 0.7865$

5. (a) $0.784 \leq \pi \leq 0.816$

   (b) $0.5099 \leq \pi \leq 0.5498$

   (c) Many more people think that emails are easier to misinterpret.

6. $0.1308 \leq \pi \leq 0.2192$

References

1. Berenson, M. L., D. M. Levine, and T. C. Krehbiel. *Basic Business Statistics: Concepts and Applications, Eleventh Edition*. Upper Saddle River, NJ: Prentice Hall, 2009.

2. Cochran, W. G. *Sampling Techniques, Third Edition*. New York: Wiley, 1977.

3. Levine, D. M., T. C. Krehbiel, and M. L. Berenson. *Business Statistics: A First Course, Fifth Edition*. Upper Saddle River, NJ: Prentice Hall, 2010.

4. Levine, D. M., D. Stephan, T. C. Krehbiel, and M. L. Berenson. *Statistics for Managers Using Microsoft Excel, Fifth Edition*. Upper Saddle River, NJ: Prentice Hall, 2008.

5. Levine, D. M., P. P. Ramsey, and R. K. Smidt, *Applied Statistics for Engineers and Scientists Using Microsoft Excel and Minitab*. Upper Saddle River, NJ: Prentice Hall, 2001.

6. Microsoft Excel 2007. Redmond, WA: Microsoft Corporation, 2006.

Fundamentals of Hypothesis Testing

7.1 The Null and Alternative Hypotheses

7.2 Hypothesis Testing Issues

7.3 Decision-Making Risks

7.4 Performing Hypothesis Testing

7.5 Types of Hypothesis Tests

One-Minute Summary

Test Yourself

Science progresses by first stating tentative explanations, or hypotheses, about natural phenomena and then by proving (or disproving) those hypotheses through investigation and testing. Statisticians have adapted this scientific method by developing an inferential method called **hypothesis testing** that evaluates a claim made about the value of a population parameter by using a sample statistic. In this chapter, you learn the basic concepts and principles of hypothesis testing and the statistical assumptions necessary for performing hypothesis testing.

7.1 The Null and Alternative Hypotheses

Unlike the broader hypothesis testing of science, statistical hypothesis testing always involves evaluating a claim made about the value of a population parameter. This claim is stated as a pair of statements: the null hypothesis and the alternative hypothesis.

Null Hypothesis

CONCEPT   The statement that a population parameter is equal to a specific value or that the population parameters from two or more groups are equal.

EXAMPLES   "The population mean time to answer customer complaints was 4 minutes last year," "the mean height for women is the same as the mean height for men," "the proportion of food orders filled correctly for drive-through customers is the same as the proportion of food orders filled correctly for sit-down customers."

important point ✏

INTERPRETATION   The null hypothesis always expresses an equality and is always paired with another statement, the alternative hypothesis. A null hypothesis is considered true until evidence indicates otherwise. If you can conclude that the null hypothesis is false, then the alternative hypothesis must be true.

You use the symbol H_0 to identify the null hypothesis and write a null hypothesis using an equal sign and the symbol for the population parameter, as in $H_0: \mu = 4$ or $H_0: \mu_1 = \mu_2$ or $H_0: \pi_1 = \pi_2$. (Remember that in statistics, the symbol π represents the population proportion and not the ratio of the circumference to the diameter of a circle, as the symbol represents in geometry.)

Alternative Hypothesis

CONCEPT   The statement paired with a null hypothesis that is mutually exclusive to the null hypothesis.

EXAMPLES   "The population mean for the time to answer customer complaints was not 4 minutes last year" (which would be paired with the example for the null hypothesis in the preceding section); "the mean height for women is not the same as the mean height for men" (paired with the second example for the null hypothesis); "the proportion of food orders filled correctly for drive-through customers is not the same as the proportion of food orders filled correctly for sit-down customers" (paired with the third example for the null hypothesis).

INTERPRETATION   The alternative hypothesis is typically the idea you are studying concerning your data. The alternative hypothesis always expresses an inequality, either between a population parameter and a specific value or between two or more population parameters and is always paired with the null hypothesis. You use the symbol H_1 to identify the alternative hypothesis and write an alternative hypothesis using either a not-equal sign or a less than or greater than sign, along with the symbol for the population parameter, as in $H_1: \mu \neq 4$ or $H_1: \mu_1 \neq \mu_2$ or $H_0: \pi_1 \neq \pi_2$.

The alternative hypothesis represents the conclusion reached by rejecting the null hypothesis. You reject the null hypothesis if evidence from the sample statistic indicates that the null hypothesis is unlikely to be true. However, if you cannot reject the null hypothesis, you cannot claim to have proven the null hypothesis. Failure to reject the null hypothesis means (only) that you have failed to prove the alternative hypothesis.

7.2 Hypothesis Testing Issues

In hypothesis testing, you use the sample statistic to estimate the population parameter stated in the null hypothesis. For example, to evaluate the null hypothesis "the population mean time to answer customer complaints was 4 minutes last year," you would use the sample mean time to estimate the population mean time. As Chapter 6, "Sampling Distributions and Confidence Intervals," establishes, a sample statistic is unlikely to be identical to its corresponding population parameter, and in that chapter you learned to construct an interval estimate for the parameter based on the statistic.

If the sample statistic is not the same as the population parameter, as it almost never is, the issue of whether to reject the null hypothesis involves deciding how different the sample statistic is from its corresponding population parameter. (In the case of two groups, the issue can be expressed, under certain conditions, as deciding how different the sample statistics of each group are to each other.)

Without a rigorous procedure that includes a clear operational definition of a difference, you would find it hard to decide on a consistent basis whether a null hypothesis is false and, therefore, whether to reject or not reject the null hypothesis. Statistical hypothesis-testing methods provide such definitions and enable you to restate the decision-making process as the probability of computing a given sample statistic, if the null hypothesis were true through the use of a test statistic and a risk factor.

Test Statistic

CONCEPT   The value based on the sample statistic and the sampling distribution for the sample statistic.

EXAMPLES   Test statistic for the difference between two sample means (Chapter 8), test statistic for the difference between two sample proportions (Chapter 8), test statistic for the difference between the means of more than two groups (Chapter 9), test statistic for the slope (Chapter 10).

INTERPRETATION   If you are testing whether the mean of a population was equal to a specific value, the sample statistic is the sample mean. The test statistic is based on the difference between the sample mean and the value of the population mean stated in the null hypothesis. This test statistic follows a statistical distribution called the *t* distribution that is discussed in Sections 8.2 and 8.3.

If you are testing whether the mean of population one is equal to the mean of population two, the sample statistic is the difference between the mean in sample one and the mean in sample two. The test statistic is based on the difference between the mean in sample one and the mean in sample two. This test statistic also follows the *t* distribution.

The sampling distribution of the test statistic is divided into two regions, a **region of rejection** (also known as the critical region) and a **region of nonrejection**. If the test statistic falls into the region of nonrejection, the null hypothesis is not rejected.

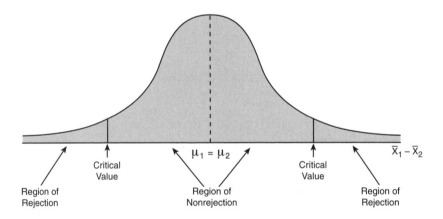

The region of rejection contains the values of the test statistic that are unlikely to occur if the null hypothesis is true. If the null hypothesis is false, these values are likely to occur. Therefore, if you observe a value of the test statistic that falls into the rejection region, you reject the null hypothesis, because that value is unlikely if the null hypothesis is true.

To make a decision concerning the null hypothesis, you first determine the critical value of the test statistic that separates the nonrejection region from the rejection region. You determine the critical value by using the appropriate sampling distribution and deciding on the risk you are willing to take of rejecting the null hypothesis when it is true.

Practical Significance Versus Statistical Significance

Another issue in hypothesis testing concerns the distinction between a statistically significant difference and a practical significant difference. Given a large enough sample size, it is always possible to detect a statistically significant difference. This is because no two things in nature are exactly equal. So, with a large enough sample size, you can always detect the natural difference between two populations. You need to be aware of the real-world practical implications of the statistical significance.

7.3 Decision-Making Risks

In hypothesis testing, you always face the possibility that either you will wrongly reject the null hypothesis or wrongly not reject the null hypothesis. These possibilities are called type I and type II errors, respectively.

Type I Error

CONCEPT   The error that occurs if the null hypothesis H_0 is rejected when it is true and should not be rejected.

INTERPRETATION   The risk, or probability, of a type I error occurring is identified by the Greek lowercase alpha, α. Alpha is also known as the level of significance of the statistical test. Traditionally, you control the probability of a type I error by deciding the risk level α you are willing to tolerate of rejecting the null hypothesis when it is true. Because you specify the level of significance before performing the hypothesis test, the risk of committing a type I error, α, is directly under your control. The most common α values are 0.01, 0.05, and 0.10, and researchers traditionally select a value of 0.05 or smaller.

When you specify the value for α, you determine the rejection region, and using the appropriate sampling distribution, the critical value or values that divide the rejection and nonrejection regions are determined.

Type II Error

CONCEPT   The error that occurs if the null hypothesis H_0 is not rejected when it is false and should be rejected.

INTERPRETATION   The risk, or probability, of a type II error occurring is identified by the Greek lowercase beta, β. The probability of a type II error depends on the size of the difference between the value of the population parameter stated in the null hypothesis and the actual population value. Unlike the type I error, the type II error is not directly established by you. Because large differences are easier to find, as the difference between the value of the population parameter stated in the null hypothesis and its corresponding population parameter increases, the probability of a type II error decreases. Therefore, if the difference between the value of the population parameter stated in the null hypothesis and the corresponding parameter is small, the probability of a type II error will be large.

The arithmetic complement of beta, $1 - \beta$, is known as the **power of the test** and represents the probability of rejecting the null hypothesis when it is false and should be rejected.

Risk Trade-Off

The types of errors and their associated risks are summarized in Table 7.1. The probabilities of the two types of errors have an inverse relationship. When you decrease α, you always increase β and when you decrease β, you always increase α.

TABLE 7.1

Risks and Decisions in Hypothesis Testing

		Actual Situation	
		H_0 **True**	H_0 **False**
Statistical Decision	Do not reject H_0	Correct decision Confidence $= 1 - \alpha$	Type II error $P(\text{Type II error}) = \beta$
	Reject H_0	Type I error $P(\text{Type I error}) = \alpha$	Correct decision Power $= 1 - \beta$

One way in which you can lower β without affecting the value of α is to increase the size of the sample. Larger sample sizes generally permit you to detect even very small differences between the hypothesized and actual values of the population parameter. For a given level of α, increasing the sample size will decrease β and therefore increase the power of the test to detect that the null hypothesis H_0 is false.

In establishing a value for α, you need to consider the negative consequences of a type I error. If these consequences are substantial, you can set $\alpha = 0.01$ instead of 0.05 and tolerate the greater β that results. If the negative consequences of a type II error most concern you, you can select a larger value for α (for example, 0.05 rather than 0.01) and benefit from the lower β that you will have.

7.4 Performing Hypothesis Testing

When you perform a hypothesis test, you should follow the steps of hypothesis testing in this order:

1. State the null hypothesis, H_0, and the alternative hypothesis, H_1.

2. Evaluate the risks of making type I and II errors, and choose the level of significance, α, and the sample size as appropriate.

3. Determine the appropriate test statistic and sampling distribution to use and identify the critical values that divide the rejection and nonrejection regions.

4. Collect the data, calculate the appropriate test statistic, and determine whether the test statistic has fallen into the rejection or the nonrejection region.

5. Make the proper statistical inference. Reject the null hypothesis if the test statistic falls into the rejection region. Do not reject the null hypothesis if the test statistic falls into the nonrejection region.

The *p*-Value Approach to Hypothesis Testing

Most modern statistical software, including the functions found in spreadsheet programs and calculators, can calculate the probability value known as the *p*-value that you can use as a second way of determining whether to reject the null hypothesis.

p-Value

CONCEPT   The probability of computing a test statistic equal to or more extreme than the sample results, given that the null hypothesis H_0 is true.

INTERPRETATION   The *p*-value is the smallest level at which H_0 can be rejected for a given set of data. You can consider the *p*-value the actual risk of having a type I error for a given set of data. Using *p*-values, the decision rules for rejecting the null hypothesis are

- If the *p*-value is greater than or equal to α, do not reject the null hypothesis.

- If the *p*-value is less than α, reject the null hypothesis.

- Many people confuse this rule, mistakenly believing that a high *p*-value is reason for rejection. You can avoid this confusion by remembering the following saying:

"If the *p*-value is low, then H_0 must go."

In practice, most researchers today use p-values for several reasons, including efficiency of the presentation of results. The p-value is also known as the **observed level of significance**. When using p-values, you can restate the steps of hypothesis testing as follows:

1. State the null hypothesis, H_0, and the alternative hypothesis, H_1.

2. Evaluate the risks of making type I and II errors, and choose the level of significance, α, and the sample size as appropriate.

3. Collect the data and calculate the sample value of the appropriate test statistic.

4. Calculate the p-value based on the test statistic and compare the p-value to α.

5. Make the proper statistical inference. Reject the null hypothesis if the p-value is less than α. Do not reject the null hypothesis if the p-value is greater than or equal to α.

7.5 Types of Hypothesis Tests

Your choice of which statistical test to use when performing hypothesis testing is influenced by the following factors:

- Number of groups of data: one, two, or more than two
- Relationship stated in alternative hypothesis H_1: not equal to or inequality (less than, greater than)
- Type of variable (population parameter): numerical (mean) or categorical (proportion)

Number of Groups

One group of hypothesis tests, more formally known as one-sample tests, are of limited practical use, because if you are interested in examining the value of a population parameter, you can usually use one of the confidence interval estimate methods of Chapter 6. Two-sample tests, examining the differences between two groups, have been the focus of this chapter and can be found in the WORKED-OUT PROBLEMS of Sections 8.1 through 8.3. Tests for more than two groups are discussed in Chapter 9.

Relationship Stated in Alternative Hypothesis H_1

Alternative hypotheses can be stated either using the not-equal sign, as in, $H_1: \mu_1 \neq \mu_2$; or by using an inequality, such as $H_1: \mu_1 > \mu_2$. You use a **two-tail test** for alternative hypotheses that use the not-equal sign and use a **one-tail test** for alternative hypotheses that contain an inequality.

One-tail and two-tail test procedures are very similar and differ mainly in the way they use critical values to determine the region of rejection. Throughout this book, two-tail hypothesis tests are featured. One-tail tests are not further discussed in this book, although WORKED-OUT PROBLEM 8 of Chapter 8 on page 171 illustrates one possible use for such tests.

Type of Variable

The type of variable, numerical or categorical, also influences the choice of hypothesis test used. For a numerical variable, the test might examine the population mean or the differences among the means, if two or more groups are used. For a categorical variable, the test might examine the population proportion or the differences among the population proportions if two or more groups are used. Tests involving two groups for each type of variable can be found in the WORKED-OUT PROBLEMS of Sections 8.1 through 8.3. Tests involving more than two groups for each type of variable are featured in Chapter 9.

One–Minute Summary

Hypotheses

- Null hypothesis
- Alternative hypothesis

Types of errors

- Type I error
- Type II error

Hypothesis testing approach

- Test statistic
- p-value

Hypothesis test relationship

- One-tail test
- Two-tail test

Test Yourself

1. A type II error is committed when:
    (a) you reject a null hypothesis that is true
    (b) you don't reject a null hypothesis that is true
    (c) you reject a null hypothesis that is false
    (d) you don't reject a null hypothesis that is false

2. A type I error is committed when:
    (a) you reject a null hypothesis that is true
    (b) you don't reject a null hypothesis that is true
    (c) you reject a null hypothesis that is false
    (d) you don't reject a null hypothesis that is false

3. Which of the following is an appropriate null hypothesis?
   (a) The difference between the means of two populations is equal to 0.
   (b) The difference between the means of two populations is not equal to 0.
   (c) The difference between the means of two populations is less than 0.
   (d) The difference between the means of two populations is greater than 0.

4. Which of the following is not an appropriate alternative hypothesis?
   (a) The difference between the means of two populations is equal to 0.
   (b) The difference between the means of two populations is not equal to 0.
   (c) The difference between the means of two populations is less than 0.
   (d) The difference between the means of two populations is greater than 0.

5. The power of a test is the probability of:
   (a) rejecting a null hypothesis that is true
   (b) not rejecting a null hypothesis that is true
   (c) rejecting a null hypothesis that is false
   (d) not rejecting a null hypothesis that is false

6. If the p-value is less than α in a two-tail test:
   (a) The null hypothesis should not be rejected.
   (b) The null hypothesis should be rejected.
   (c) A one-tail test should be used.
   (d) No conclusion can be reached.

7. A test of hypothesis has a type I error probability (α) of 0.01. Therefore:
   (a) If the null hypothesis is true, you don't reject it 1% of the time.
   (b) If the null hypothesis is true, you reject it 1% of the time.
   (c) If the null hypothesis is false, you don't reject it 1% of the time.
   (d) If the null hypothesis is false, you reject it 1% of the time.

8. Which of the following statements is not true about the level of significance in a hypothesis test?
   (a) The larger the level of significance, the more likely you are to reject the null hypothesis.

(b) The level of significance is the maximum risk you are willing to accept in making a type I error.

(c) The significance level is also called the α level.

(d) The significance level is another name for a type II error.

9. If you reject the null hypothesis when it is false, then you have committed:

(a) a type II error

(b) a type I error

(c) no error

(d) a type I and type II error

10. The probability of a type _____ error is also called "the level of significance."

11. The probability of a type I error is represented by the symbol _____.

12. The value that separates a rejection region from a non-rejection region is called the _____.

13. Which of the following is an appropriate null hypothesis?

(a) The mean of a population is equal to 100.

(b) The mean of a sample is equal to 50.

(c) The mean of a population is greater than 100.

(d) All of the above.

14. Which of the following is an appropriate alternative hypothesis?

(a) The mean of a population is equal to 100.

(b) The mean of a sample is equal to 50.

(c) The mean of a population is greater than 100.

(d) All of the above.

Answer True or False:

15. For a given level of significance, if the sample size is increased, the power of the test will increase.

16. For a given level of significance, if the sample size is increased, the probability of committing a type I error will increase.

17. The statement of the null hypothesis always contains an equality.

18. The larger the *p*-value, the more likely you are to reject the null hypothesis.

19. The statement of the alternative hypothesis always contains an equality.

20. The smaller the *p*-value, the more likely you are to reject the null hypothesis.

Answers to Test Yourself

1. d		11. α	
2. a		12. critical value	
3. a		13. a	
4. a		14. c	
5. c		15. True	
6. b		16. False	
7. b		17. True	
8. d		18. False	
9. c		19. False	
10. I		20. True	

References

1. Berenson, M. L., D. M. Levine, and T. C. Krehbiel. *Basic Business Statistics: Concepts and Applications, Eleventh Edition.* Upper Saddle River, NJ: Prentice Hall, 2009.

2. Cochran, W. G. *Sampling Techniques, Third Edition.* New York: Wiley, 1977.

3. Levine, D. M., T. C. Krehbiel, and M. L. Berenson. *Business Statistics: A First Course, Fifth Edition.* Upper Saddle River, NJ: Prentice Hall, 2010.

4. Levine, D. M., D. Stephan, T. C. Krehbiel, and M. L. Berenson. *Statistics for Managers Using Microsoft Excel, Fifth Edition.* Upper Saddle River, NJ: Prentice Hall, 2008.

CHAPTER

8

Hypothesis Testing:
Z and *t* Tests

8.1   Testing for the Difference Between Two
      Proportions
8.2   Testing for the Difference Between the Means of
      Two Independent Groups
8.3   The Paired *t* Test
      Important Equations
      One-Minute Summary
      Test Yourself

In Chapter 7, you learned the fundamentals of hypothesis testing. This chapter discusses hypothesis tests that involve two groups, more formally known as two-sample tests. You will learn to use:

- The hypothesis test that examines the differences between the proportions of two groups

- The hypothesis test that examines the differences between the means of two groups

You will also learn how to evaluate the statistical assumptions about your data that need to be true in order to use these tests and what to do if the assumptions do not hold.

8.1 Testing for the Difference Between Two Proportions

CONCEPT   Hypothesis test that analyzes differences between two groups by examining the proportion of items in each group that are in a particular category.

INTERPRETATION   For this test, the test statistic is based on the difference in the sample proportion of the two groups. Both the sample proportion of occurrences in group 1 and the sample proportion of occurrences in group 2 are needed to perform this test. With a sufficient sample size in each group, the sampling distribution of the difference between the two proportions approximately follows a normal distribution (see Section 5.3).

WORKED-OUT PROBLEM 1   The following two-way table summarizes the results of a study that investigated the manufacture of silicon wafers. This table presents the joint responses of whether a specific wafer was "good" or "bad" and whether that wafer had particles present on it.

Counts of Particles Found Cross-Classified by Wafer Condition

		Wafer Condition		Total
		Good	Bad	
Particles Present	Yes	14	36	50
	No	320	80	400
	Total	334	116	450

Of 334 wafers that were classified as good, 320 had no particles found on the dye that produced the wafer. Of 116 wafers that were classified as bad, 80 had no particles found on the dye that produced the wafer. You want to determine whether the proportion of wafers with no particles is the same for good and bad wafers using a level of significance of $\alpha = 0.05$.

For these data, the proportion of good wafers without particles is

$$\frac{320}{334} = 0.9581$$

and the proportion of bad wafers without particles is

$$\frac{80}{116} = 0.6897$$

Because the number of good and bad wafers that have no particles (320 and 80) is large, as is the number of good and bad wafers that have particles (14 and 36), the sampling distribution for the difference between the two proportions is approximately normally distributed. The null and alternative hypotheses are as follows:

$$H_0: \pi_1 = \pi_2 \text{ (No difference between the proportions}$$
$$\text{for the "good" and "bad" groups.)}$$

$H_1: \pi_1 \neq \pi_2$ (There is a difference between
the proportions for the two groups.)

Spreadsheet and calculator results for the manufacturing study are as follows:

	A	B
1	Z Test for Differences in Two Proportions	
2		
3	Data	
4	Hypothesized Difference	0
5	Level of Significance	0.05
6	Group 1	
7	Number of Items of Interest	320
8	Sample Size	334
9	Group 2	
10	Number of Items of Interest	80
11	Sample Size	116
12		
13	Intermediate Calculations	
14	Group 1 Proportion	0.9581
15	Group 2 Proportion	0.6897
16	Difference in Two Proportions	0.2684
17	Average Proportion	0.8889
18	Z Test Statistic	7.9254
19		
20	Two-Tail Test	
21	Lower Critical Value	-1.9600
22	Upper Critical Value	1.9600
23	p-Value	0.0000
24	Reject the null hypothesis	

```
2-PropZTest
 P1≠P2
 z=7.92542153
 p=2.296343ᴇ-15
 p̂1=.9580838323
 p̂2=.6896551724
↓p̂=.8888888889
```

Using the critical value approach with a level of significance of 0.05, the
lower tail area is 0.025, and the upper tail area is 0.025. Using the cumulative normal distribution table (Table C.1), the lower critical value of 0.025
corresponds to a Z value of –1.96, and an upper critical value of 0.025
(cumulative area of 0.975) corresponds to a Z value of +1.96, as shown in
the following diagram:

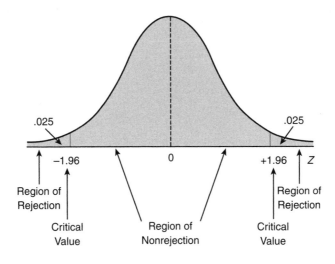

.025   .025

        −1.96   0   +1.96   Z

Region of   Region of
Rejection   Rejection

        Critical   Region of   Critical
        Value   Nonrejection   Value

Given these rejection regions, you will reject H_0 if $Z < -1.96$ or if $Z > +1.96$; otherwise you will not reject H_0. Microsoft Excel worksheet (and TI calculator) calculations shown earlier determine that the Z test statistic is 7.93. Because $Z = 7.93$ is greater than the upper critical value of $+1.96$, you reject the null hypothesis. You conclude that evidence exists of a difference in the proportion of good and bad wafers that have no particles.

WORKED-OUT PROBLEM 2   You decide to use the *p*-value approach to hypothesis testing. Calculations done in a spreadsheet and the TI calculator (see the figures on page 157) determine that the *p*-value is 0.0000. This means that the probability of obtaining a Z value greater than 7.93 is virtually zero (0.0000). Because the *p*-value is less than the level of significance $\alpha = 0.05$, you reject the null hypothesis. You conclude that evidence exists of a difference in the proportion of good and bad wafers that have no particles. Clearly, the good wafers are much more likely to have no particles than the bad wafers.

calculator keys

Z Test for the Difference Between Two Proportions

Press [STAT][◀] (to display the Tests menu) and then select **6:2-PropZTest** and press [ENTER] to display the 2-PropZTest screen. In this screen, enter values for the group 1 number of successes and sample size (**x1** and **n1**) and the group 2 number of successes and sample size (**x2** and **n2**). Select the first alternative hypothesis choice and then select Calculate and press [ENTER]. (The level of significance is preset to $\alpha = 0.05$ and cannot be changed.) If the *p*-value is a very small value, it might appear as a number in exponential notation such as 5.7009126E-7. Treat such values as being equivalent to zero.

WORKED-OUT PROBLEM 3   You need to analyze the results of a famous health care experiment that investigated the effectiveness of aspirin in the reduction of the incidence of heart attacks, using a level of significance of $\alpha = 0.05$. In this experiment, 22,071 male U.S. physicians were randomly assigned to either a group that was given one 325 mg buffered aspirin tablet every other day or a group that was given a placebo (a pill that contained no active ingredients). Of 11,037 physicians taking aspirin, 104 suffered heart attacks during the five-year period of the study. Of 11,034 physicians who were assigned to a group that took a placebo every other day, 189 suffered heart attacks during the five-year period of the study. You summarize these results as follows:

Results Classified by Whether the Physician Took Aspirin

		Study Group		
		Aspirin	**Placebo**	**Totals**
Results	Heart attack	104	189	293
	No heart attack	10,933	10,845	21,778
	Totals	11,037	11,034	22,071

You establish the null and alternative hypotheses as

H_0: $\pi_1 = \pi_2$ (No difference exists in the proportion of heart attacks between the group that was given aspirin and the group that was given the placebo.)

H_1: $\pi_1 \neq \pi_2$ (A difference exists in the proportion of heart attacks between the two groups.)

Spreadsheet and calculator results for the health care experiment data are as follows:

	A	B
1	Z Test for Differences in Two Proportions	
2		
3	Data	
4	Hypothesized Difference	0
5	Level of Significance	0.05
6	Group 1	
7	Number of Items of Interest	104
8	Sample Size	11037
9	Group 2	
10	Number of Items of Interest	189
11	Sample Size	11034
12		
13	Intermediate Calculations	
14	Group 1 Proportion	0.0094
15	Group 2 Proportion	0.0171
16	Difference in Two Proportions	-0.0077
17	Average Proportion	0.0133
18	Z Test Statistic	-5.0014
19		
20	Two-Tail Test	
21	Lower Critical Value	-1.9600
22	Upper Critical Value	1.9600
23	*p*-Value	0.0000
24	Reject the null hypothesis	

```
2-ProFZTest
 P1≠P2
 z=-5.001388204
 P=5.7009126E-7
 P̂1=.0094228504
 P̂2=.0171288744
↓P̂=.0132753387
```

These results show that the p-value is 0 and the value of the test statistic is Z = 5.00. (The calculator result for the p-value is 5.7E-7, a tiny decimal fraction that you can consider equivalent to zero.) This means that the chance of obtaining a Z value greater than 5.00 is virtually zero. Because the p-value (0) is less than the level of significance (0.05), you reject the null hypothesis and accept the alternative hypothesis that a difference exists in the proportion of heart attacks between the two groups. (Using the critical value approach, because Z = 5.00 is greater than the upper critical value of +1.96 at the 0.05 level of significance, you can reject the null hypothesis.)

You conclude that evidence exists of a difference in the proportion of doctors who have had heart attacks between those who took the aspirin and those who did not take the aspirin. The study group who took the aspirin had a significantly lower proportion of heart attacks over the study period.

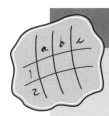

spreadsheet solution

Z Test for the Difference in Two Proportions

Chapter 8 Z Two Proportions performs the Z test for the difference in two proportions based on the values for the hypothesized difference, level of significance, and number of successes and sample size for each group that you enter. (Already entered into the spreadsheet as an example are the values used in WORKED-OUT PROBLEMs 1 and 2.)

Spreadsheet Tips FT7 and FT8 in Appendix D explain the spreadsheet functions used to calculate the critical values and the p-value.

equation blackboard (optional)

interested in math?

WORKED-OUT PROBLEMs 1–3 use the Z test for the difference between two proportions. You need the subscripted symbols for the number of successes, X, the sample sizes, n_1 and n_2, sample proportions, p_1 and p_2, and population proportions, π_1 and π_2, as well as the symbol for the pooled estimate of the population proportion, $\bar{p}$, to calculate the Z test statistic.

To write the Z test statistic equation, you first define the symbols for the pooled estimate of the population proportion and the sample proportions for the two groups:

$$\bar{p} = \frac{X_1 + X_2}{n_1 + n_2} \quad p_1 = \frac{X_1}{n_1} \quad p_2 = \frac{X_2}{n_2}$$

Next, you use $\bar{p}$, p_1, and p_2 along with the symbols for the sample sizes and population proportion to form the equation for the Z test for the difference between two proportions:

$$Z = \frac{(p_1 - p_2) - (\pi_1 - \pi_2)}{\sqrt{\bar{p}(1 - \bar{p})\left(\dfrac{1}{n_1} + \dfrac{1}{n_2}\right)}}$$

As an example, the calculations for determining the Z test statistic for WORKED-OUT PROBLEM 1 concerning the manufacture of silicon wafers are as follows:

$$p_1 = \frac{X_1}{n_1} = \frac{320}{334} = 0.9581 \quad p_2 = \frac{X_2}{n_2} = \frac{80}{116} = 0.6897$$

and

$$\bar{p} = \frac{X_1 + X_2}{n_1 + n_2} = \frac{320 + 80}{334 + 116} = \frac{400}{450} = 0.8889$$

so that

$$Z = \frac{(0.9581 - 0.6897) - (0)}{\sqrt{0.8889(1 - 0.8889)\left(\dfrac{1}{334} + \dfrac{1}{116}\right)}}$$

$$= \frac{0.2684}{\sqrt{0.8889(1 - 0.8889)(0.0116)}}$$

$$= \frac{0.2684}{\sqrt{(0.09876)(0.0116)}}$$

$$= \frac{0.2684}{\sqrt{0.0011456}}$$

$$= \frac{0.2684}{0.03385} = +7.93$$

With $\alpha = 0.05$, you reject H_0 if $Z < -1.96$ or if $Z > +1.96$; otherwise, do not reject H_0. Because $Z = 7.93$ is greater than the upper critical value of $+1.96$, you reject the null hypothesis.

8.2 Testing for the Difference Between the Means of Two Independent Groups

CONCEPT   Hypothesis test that analyzes differences between two groups by determining whether a significant difference exists in the population means (a numerical parameter) of two populations or groups.

INTERPRETATION   Statisticians distinguish between using two independent groups and using two related groups when performing this type of hypothesis test. With related groups, the observations are either matched according to a relevant characteristic or repeated measurements of the same items are taken. For studies involving two independent groups, the most common test of hypothesis used is the pooled-variance *t* test.

Pooled-Variance *t* Test

CONCEPT   The hypothesis test for the difference between the population means of two independent groups that combines or "pools" the sample variance of each group into one estimate of the variance common in the two groups.

INTERPRETATION   For this test, the test statistic is based on the difference in the sample means of the two groups, and the sampling distribution for the difference in the two sample means approximately follows the *t* distribution.

In a pooled variance *t* test, the null hypothesis of no difference in the means of two independent populations is

$$H_0: \mu_1 = \mu_2 \text{ (The two population means are equal.)}$$

and the alternative hypothesis is

$$H_1: \mu_1 \neq \mu_2 \text{ (The two population means are not equal.)}$$

WORKED-OUT PROBLEM 4   You want to determine whether the cost of a restaurant meal in a major city differs from the cost of a similar meal in the suburbs outside the city. You collect data about the cost of a meal per person from a sample of 50 city restaurants and 50 suburban restaurants as follows:

RestCost

City Cost Data									
13	21	22	22	24	25	26	26	26	26
30	32	33	34	34	35	35	35	35	36
37	37	39	39	39	40	41	41	41	42
43	44	45	46	50	50	51	51	53	53
53	55	57	61	62	62	62	66	68	75

Suburban Cost Data									
21	22	25	25	26	26	27	27	28	28
28	29	31	32	32	35	35	36	37	37
37	38	38	38	39	40	40	41	41	41
42	42	43	44	47	47	47	48	50	50
50	50	50	51	52	53	58	62	65	67

The Excel Analysis ToolPak results for the cost of the restaurant meals are as follows:

	A	B	C
1	t-Test: Two-Sample Assuming Equal Variances		
2			
3		City Cost	Suburban Cost
4	Mean	41.46	39.96
5	Variance	192.9065	123.9984
6	Observations	50	50
7	Pooled Variance	158.4524	
8	Hypothesized Mean Difference	0	
9	df	98	
10	t Stat	0.5958	
11	P(T<=t) one-tail	0.2763	
12	t Critical one-tail	1.6606	
13	P(T<=t) two-tail	0.5527	
14	t Critical two-tail	1.9845	

These results show that the t statistic (labeled as **t Stat** in the results) is 0.5958 and the p-value (labeled as **P(T< = t) two-tail**) is 0.5527. Because $t =$ 0.5958 < 1.9845 or because the p-value, 0.5527, is greater than $\alpha = 0.05$, you do not reject the null hypothesis. You conclude that the chance of obtaining a t value greater than 0.5958 is large (0.5527) and therefore have insufficient evidence of a significant difference in the cost of a restaurant meal in the city (sample mean is $41.46) and a restaurant meal in the suburbs outside the city (sample mean of $39.96).

calculator keys

Pooled Variance *t* Test for the Differences in Two Means

Press [STAT][◀] (to display the TESTS menu) and then select 4:2-SampTTest and press [ENTER] to display the 2-SampTTest screen. Then proceed with the appropriate procedure:

- **When using sample data.** Select **Data** as the **Inpt** type and press [ENTER]. Enter the names of the list variables for each sample and make sure **Freq1** and **Freq2** are both set to **1**. Select the first alternative hypothesis, press [ENTER], and for **Pooled**, select **Yes** and then [ENTER]. Press [▼] and then select **Calculate** and press [ENTER].

- **When using sample statistics.** Select **Stats** as the **Inpt** type and press [ENTER]. Enter the sample mean, sample standard deviation, and sample size of group 1, followed by those statistics for group 2. Select the first alternative hypothesis and press [ENTER]. Press [▼] and for Pooled, select Yes and then [ENTER]. Press [▼] and then select **Calculate** and press [ENTER].

For these tests, the level of significance is preset to $\alpha = 0.05$ and cannot be changed. If the *p*-value is a very small value, it can appear as a number in exponential notation that you should consider to be equivalent to zero.

WORKED-OUT PROBLEM 5   You want to determine at a level of significance of $\alpha = 0.05$ whether the mean payment made by online customers of a website differs according to two methods of payment. You obtained the following statistics based on a random sample of 50 transactions.

	Method 1	Method 2
Sample Size	22.0	28.0
Sample Mean	30.37	23.17
Sample Standard Deviation	12.006	7.098

spreadsheet solution

Pooled Variance *t* Test for the Differences in Two Means

Test with sample data. Chapter 8 Pooled-Variance t ATP (shown on page 161) contains the results of using the Analysis ToolPak **t-Test: Two-Sample Assuming Equal Variances** procedure to perform a pooled-variance *t* test using the restaurant meal cost sample data of WORKED-OUT PROBLEM 4.

Spreadsheet Tip ATT3 in Appendix D explains how to use the **t-Test: Two-Sample Assuming Equal Variances** procedure.

Test with sample statistics. Chapter 8 Pooled-Variance t performs a pooled variance *t* test based on the values for the hypothesized difference, the level of significance, and the sample size, sample mean, and sample standard deviation for each group that you enter. (Already entered as an example are the values used in WORKED-OUT PROBLEM 5.)

Spreadsheet Tips FT6 and FT9 in Appendix D explain the spreadsheet functions used to calculate the critical values and the *p*-value.

Spreadsheet and calculator results for this study are as follows:

A	B
1 Pooled-Variance *t* Test	
2	
3 Data	
4 Hypothesized Difference	0
5 Level of Significance	0.05
6 Group 1 Sample	
7 Sample Size	22
8 Sample Mean	30.37
9 Sample Standard Deviation	12.006
10 Group 2 Sample	
11 Sample Size	28
12 Sample Mean	23.17
13 Sample Standard Deviation	7.098
14	
15 Intermediate Calculations	
16 Population 1 Sample Degrees of Freedom	21
17 Population 2 Sample Degrees of Freedom	27
18 Total Degrees of Freedom	48
19 Pooled Variance	91.4027
20 Difference in Sample Means	7.2000
21 *t* Test Statistic	2.6434
22	
23 Two-Tail Test	
24 Lower Critical Value	-2.0106
25 Upper Critical Value	2.0106
26 *p*-Value	0.0111
27 Reject the null hypothesis	

```
2-SampTTest
µ1≠µ2
t=2.643372782
p=.0110544755
df=48
x̄1=30.37
↓x̄2=23.17
```

The results show that the *t* statistic is 2.64 and the *p*-value is 0.0111. Because the *p*-value is less than $\alpha = 0.05$, you reject the null hypothesis. (Using the critical value approach, $t = 2.6434 > 2.01$, leading you to the same decision.) You conclude that the chance of obtaining a *t* value greater than 2.64 is very small (0.0111) and therefore assert that the mean payment amount is higher for method 1 (sample mean of $30.37) than for method 2 (sample mean of $23.17).

equation blackboard (optional)

interested in math?

WORKED-OUT PROBLEMS 4 and 5 use the pooled variance *t* test for the difference between the population means of two independent groups. You need the subscripted symbols for the sample means, $\overline{X}_1$ and $\overline{X}_2$, the sample sizes for each of the two groups, n_1 and n_2, and the population means, μ_1 and μ_2 along with the symbol for the pooled estimate of the variance, S_p^2, to calculate the *t* test statistic.

To write the equation for the *t* test statistic, you first define the symbols for the equation for the pooled estimate of the population variance:

$$S_p^2 = \frac{(n_1 - 1)S_1^2 + (n_2 - 1)S_2^2}{(n_1 - 1) + (n_2 - 1)}$$

You next use S_p^2 that you just defined and the symbols for the sample means, the population means, and the sample sizes to form the equation for the pooled-variance t test for the difference between two means:

$$t = \frac{(\overline{X}_1 - \overline{X}_2) - (\mu_1 - \mu_2)}{\sqrt{S_p^2\left(\dfrac{1}{n_1} + \dfrac{1}{n_2}\right)}}$$

The calculated test statistic *t* follows a *t* distribution with $n_1 + n_2 - 2$ degrees of freedom.

As an example, the calculations for determining the *t* test statistic for the restaurant cost data with $\alpha = 0.05$, are as follows:

$$S_p^2 = \frac{(n_1 - 1)S_1^2 + (n_2 - 1)S_2^2}{(n_1 - 1) + (n_2 - 1)}$$

$$= \frac{49(192.9065) + 49(123.9984)}{49 + 49} = 158.4524$$

Using 158.4524 as the value for S_p^2 in the original equation

$$t = \frac{(\bar{X}_1 - \bar{X}_2) - (\mu_1 - \mu_2)}{\sqrt{S_p^2 \left(\frac{1}{n_1} + \frac{1}{n_2} \right)}}$$

produces

$$t = \frac{(41.46 - 39.96) - 0}{\sqrt{158.4524 \left(\frac{1}{50} + \frac{1}{50} \right)}}$$

$$t = \frac{41.46 - 39.96}{\sqrt{158.4524(0.04)}}$$

$$= \frac{1.50}{\sqrt{6.3381}} = +0.5958$$

Using the $\alpha = 0.05$ level of significance, with $50 + 50 - 2 = 98$ degrees of freedom the critical value of t is 1.9845 (0.025 in the upper tail of the t distribution). Because $t = +0.5958 < 1.9845$, you don't reject H_0.

Pooled-Variance t Test Assumptions

In testing for the difference between the means, you assume that the two populations from which the two independent samples have been selected are normally distributed with equal variances. When the two populations *do* have equal variances, the pooled-variance t test is valid even if there is a moderate departure from normality, as long as the sample sizes are large.

You can check the assumption of normality by preparing a side-by-side box-and-whisker plot for the two samples. Such a plot was shown for the meal cost at city and suburban restaurants in Chapter 3, "Descriptive Statistics," on page 61. These box-and-whisker plots have some right-skewness because the tail on the right is longer than the one on the left. However, given the relatively large sample size in each of the two groups, you can conclude that any departure from the normality assumption will not seriously affect the validity of the t test. If the data in each group cannot be assumed to be from normally distributed populations, you can use a nonparametric procedure, such as the Wilcoxon rank sum test (see References 1–4) that does not depend on the assumption of normality in the two populations.

The pooled-variance t test also assumes that the population variances are equal. If this assumption cannot be made, you cannot use the pooled-variance t test. In such cases, you can use the separate-variance t test (see References 1–4).

8.3 **The Paired t Test**

CONCEPT   A hypothesis test for the difference between two groups for situations in which the data from the two groups is *related and not independent*. In this test, the variable of interest is the difference between related pairs of values in the two groups, rather than the paired values themselves.

INTERPRETATION   There are two situations in which the data values from two groups will be related and not independent.

In the first case, a researcher has paired, or matched, the values under study according to some other variable. For example, in testing whether a new drug treatment lowers blood pressure, a sample of patients could be paired according to their blood pressure at the beginning of the study. For example, if there were two patients in the study who each had a diastolic blood pressure of 140, one would be randomly assigned to the group that will take the new drug and the other to the group that will not take the new drug. Assigning patients in this manner means that the researcher will not have to worry about differences in the initial blood pressures of the patients that form the two groups. This, in turn, means that test results will better reflect the effect of the new drug being tested.

In the second case, a researcher obtains two sets of measurements from the same items or individuals. This approach is based on the theory that the same items or individuals will behave alike if treated alike. This, in turn, allows the researcher to assert that any differences between two sets of measurements are due to what is under study. For example, when performing an experiment on the effect of a diet drug, the researcher could take one measurement from each participant just prior to starting the drug and one just after the end of a specified time period.

In both cases, the variable of interest can be stated algebraically as follows:

Difference (D) = Related value in sample 1 − Related value in sample 2

With related groups and a numeric variable of interest, the null hypothesis is that no difference exists in the population means of the two related groups, and the alternative hypothesis is that there is a difference in the population means of the two related groups. Using the symbol μ_D to represent the difference between the population means, the null and alternative hypotheses can be expressed as follows:

$$H_0: \mu_D = 0$$

and

$$H_1: \mu_D \neq 0$$

To decide whether to reject the null hypothesis, you use a paired *t* test.

WORKED-OUT PROBLEM 6 You want to determine, using a level of significance of $\alpha = 0.05$, whether a difference exists between the Doppler echocardiography measurements that two different observers made working with a sample of 23 patients. The paired measurements and their differences are as follows:

Cardiac

Patient	Observer A	Observer B	Difference (*D*)
1	4.8	5.8	−1.0
2	5.6	6.1	−0.5
3	6.0	7.7	−1.7
4	6.4	7.8	−1.4
5	6.5	7.6	−1.1
6	6.6	8.1	−1.5
7	6.8	8.0	−1.2
8	7.0	8.1	−1.1
9	7.0	6.6	0.4
10	7.2	8.1	−0.9
11	7.4	9.5	−2.1
12	7.6	9.6	−2.0
13	7.7	8.5	−0.8
14	7.7	9.5	−1.8
15	8.2	9.1	−0.9
16	8.2	10.0	−1.8
17	8.3	9.1	−0.8
18	8.5	10.8	−2.3
19	9.3	11.5	−2.2
20	10.2	11.5	−1.3
21	10.4	11.2	−0.8
22	10.6	11.5	−0.9
23	11.4	12.0	−0.6

Source: Extracted from Ernst, M. L. R. Guerra, and W. R. Schucany, "Scatterplots for unordered pairs," *The American Statistician*, 1996, 50, 260–265.

Because the two observers measure the same set of patients, the two groups of measurements are related and the differences between the two groups are tested.

The Excel Analysis ToolPak results for this study are as follows:

	A	B	C
1	t-Test: Paired Two Sample for Means		
2			
3		Observer A	Observer B
4	Mean	7.8000	9.0304
5	Variance	2.8027	3.2213
6	Observations	23	23
7	Pearson Correlation	0.9346	
8	Hypothesized Mean Difference	0.0000	
9	df	22	
10	t Stat	-9.2421	
11	P(T<=t) one-tail	0.0000	
12	t Critical one-tail	1.7171	
13	P(T<=t) two-tail	0.0000	
14	t Critical two-tail	2.0739	

The results show that the t statistic (labeled **t Stat**) is -9.24 and the p-value (labeled as **P(T<=t) two-tail**) is 0.0000. Because the p-value is 0.0000 is less than $\alpha = 0.05$ (or because $t = -9.24 < -2.0739$), you reject the null hypothesis. This means that the chance of obtaining a t value less than -9.24 is virtually zero and you conclude that a difference exists between the observers.

spreadsheet solution

Paired t Test

Chapter 8 Paired t ATP (shown above) contains the results of using the Analysis ToolPak **t-Test: Paired Two Sample for Means** procedure to perform a paired t test that uses the Doppler echocardiography measurements of WORKED-OUT PROBLEM 6.

Spreadsheet Tip ATT4 in Appendix D explains how to use the **t-Test: Paired Two Sample for Means** procedure.

equation
blackboard
(optional)

interested
in
math?

WORKED-OUT PROBLEM 6 uses the equation for the paired t test. You need the symbols for the sample size, n, the difference between the population means, μ_D, the sample standard deviation, S_D, and the subscripted symbol for the differences in the paired values, D_i, all previously introduced, and the symbol for the mean difference, $\bar{D}$, to calculate the t test statistic.

To write the t test statistic equation, you first define the symbols for the equation for the mean difference, $\bar{D}$:

$$\bar{D} = \frac{\sum\limits_{i=1}^{n} D_i}{n}$$

You next use $\bar{D}$, the symbols for the sample size, and the differences in the paired values to form the equation for the sample standard deviation, S_D:

$$S_D = \sqrt{\frac{\sum\limits_{i=1}^{n} (D_i - \bar{D})^2}{n-1}}$$

Finally, you assemble $\bar{D}$ and S_D and the remaining symbols to form the equation for the paired t test for the difference between two means:

$$t = \frac{\bar{D} - \mu_D}{\dfrac{S_D}{\sqrt{n}}}$$

The test statistic t follows a t distribution with $n - 1$ degrees of freedom.

As an example, the calculations for determining the t test statistic for WORKED-OUT PROBLEM 6 concerning the difference between the Doppler echocardiography measurements with $\alpha = 0.05$, are as follows:

$$\bar{D} = \frac{\sum\limits_{i=1}^{n} D_i}{n}$$

$$= \frac{-28.3}{23} = -1.23$$

This makes $S_D = 0.638$ (calculation not shown). Substituting these values results in

(continues)

$$t = \frac{\bar{D} - \mu_D}{\frac{S_D}{\sqrt{n}}} = \frac{-1.23 - 0}{\frac{0.638}{\sqrt{23}}} = -9.24$$

Using the $\alpha = 0.05$ level of significance, with $23 - 1 = 22$ degrees of freedom, the critical value of t is -2.0739 (0.025 in the lower tail of the t distribution). Because $t = -9.24 < -2.0739$, you reject the null hypothesis H_0.

WORKED-OUT PROBLEM 7   You seek to determine, using a level of significance of $\alpha = 0.05$, whether differences exist in monthly sales between the new package design and the old package design of a laundry stain remover. The new package was test marketed over a period of one month in a sample of supermarkets in a particular city. A random sample of ten pairs of supermarkets was matched according to weekly sales volume and a set of demographic characteristics. The data collected for this study are as follows:

Supermarket

Monthly Sales of Laundry Stain Remover

Pair	New Package	Old Package	Difference
1	458	437	21
2	519	488	31
3	394	409	-15
4	632	587	45
5	768	753	15
6	348	400	-52
7	572	508	64
8	704	695	9
9	527	496	31
10	584	513	71

Because the ten pairs of supermarkets were matched, you use the paired t test. The Excel Analysis ToolPak results for this study are as follows:

	A	B	C
1	t-Test: Paired Two Sample for Means		
2			
3		New Package	Old Package
4	Mean	550.6	528.6
5	Variance	17188.27	13785.16
6	Observations	10	10
7	Pearson Correlation	0.9631	
8	Hypothesized Mean Difference	0	
9	df	9	
10	t Stat	1.9116	
11	P(T<=t) one-tail	0.0441	
12	t Critical one-tail	1.8331	
13	P(T<=t) two-tail	0.0882	
14	t Critical two-tail	2.2622	

These results show that the t statistic (labeled as **t Stat**) is 1.91 and the p-value (labeled as **P(T<t) two-tail**) is 0.0882. This means that the chance of obtaining a t value greater than 1.91 or less than -1.91 is 0.0882 or 8.82%. Because the p-value is 0.0882 is greater than $\alpha = 0.05$, you do not reject the null hypothesis. (Using the critical value approach, $t = 1.91 < 2.2622$, and you reach the same decision.) You can conclude that insufficient evidence exists of a difference between the new and old package design.

WORKED-OUT PROBLEM 8   Before the test marketing experiment, you might have wanted to determine whether the new package design produced *more* sales than the old package design and not just a difference in sales. Such a situation would be a good application of a one-tail test in which the alternative hypothesis is $H_1: \mu_D > 0$.

In such a case, you would use the one-tail p-value (or one-tail critical value). For the test market sales data, the one-tail p-value is 0.0441 (and the one-tail critical value is 1.8331). Because the p-value is less than $\alpha = 0.05$, (or because using the critical value approach, $t = 1.911$ is greater than 1.8331), you reject the null hypothesis and conclude that the mean sales from the new package design were higher than the mean sales for the old package design—a different conclusion than you made using the two-tail test.

Important Equations

Z Test for the Difference Between Two Proportions:

(8.1)
$$Z = \frac{(p_1 - p_2) - (\pi_1 - \pi_2)}{\sqrt{\bar{p}(1-\bar{p})\left(\dfrac{1}{n_1} + \dfrac{1}{n_2}\right)}}$$

Pooled Variance t Test for the Difference Between the Population Means of Two Independent Groups:

(8.2)
$$t = \frac{(\bar{X}_1 - \bar{X}_2) - (\mu_1 - \mu_2)}{\sqrt{S_p^2\left(\dfrac{1}{n_1} + \dfrac{1}{n_2}\right)}}$$

Paired t Test for the Difference Between Two Means:

(8.3)
$$t = \frac{\bar{D} - \mu_D}{\dfrac{S_D}{\sqrt{n}}}$$

One-Minute Summary

For tests for the differences between two groups, first determine whether your data are categorical or numerical.

- If your data are categorical, use the Z test for the difference between two proportions.

- If your data are numerical, determine whether you have independent or related groups:

    If you have independent groups, use the pooled variance t test for the difference between two means.

    If you have related groups, use the paired t test.

Test Yourself

Short Answers

1. The t test for the difference between the means of two independent populations assumes that the two:
    (a) Sample sizes are equal.
    (b) Sample medians are equal.
    (c) Populations are approximately normally distributed.
    (d) All of the above.

2. In testing for differences between the means of two related populations, the null hypothesis is:
    (a) $H_0: \mu_D = 2$
    (b) $H_0: \mu_D = 0$
    (c) $H_0: \mu_D < 0$
    (d) $H_0: \mu_D > 0$

3. A researcher is curious about the effect of sleep on students' test performances. He chooses 100 students and gives each student two exams. One is given after four hours' sleep and one after eight hours' sleep. The statistical test the researcher should use is the:
    (a) Z test for the difference between two proportions
    (b) Pooled-variance t test
    (c) Paired t test

4. A statistics professor wanted to test whether the grades on a statistics test were the same for her morning class and her afternoon class. For this situation, the professor should use the:
    (a) Z test for the difference between two proportions
    (b) Pooled-variance t test
    (c) Paired t test

Answer True or False:

5. The sample size in each independent sample must be the same in order to test for differences between the means of two independent populations.

6. In testing a hypothesis about the difference between two proportions, the *p*-value is computed to be 0.043. The null hypothesis should be rejected if the chosen level of significance is 0.05.

7. In testing a hypothesis about the difference between two proportions, the *p*-value is computed to be 0.034. The null hypothesis should be rejected if the chosen level of significance is 0.01.

8. In testing a hypothesis about the difference between two proportions, the Z test statistic is computed to be 2.04. The null hypothesis should be rejected if the chosen level of significance is 0.01 and a two-tail test is used.

9. The sample size in each independent sample must be the same in order to test for differences between the proportions of two independent populations.

10. When you are sampling the same individuals and taking a measurement before treatment and after treatment, you should use the paired *t* test.

11. Repeated measurements from the same individuals are an example of data collected from two related populations.

12. The pooled-variance *t* test assumes that the population variances in the two independent groups are equal.

13. In testing a null hypothesis about the difference between two proportions, the Z test statistic is computed to be 2.04. The *p*-value is 0.0207.

14. You can use a pie chart to evaluate whether the assumption of normally distributed populations in the pooled-variance *t* test has been violated.

15. If the assumption of normally distributed populations in the pooled-variance *t* test has been violated, you should use an alternative procedure such as the nonparametric Wilcoxon rank sum test.

Answers to Test Yourself Short Answers

1. c

2. b

3. c

4. b

5. False

6. True

7. False

8. False

9. False

10. True

11. True

12. True

13. False

14. False

15. True

Problems

1. Technology has led to the rise of extreme workers who are on the job 60 hours or more per week. One of the reasons cited by employees as to why they worked long hours was that they loved their job because it is stimulating/challenging/provides an adrenaline rush (extracted from S. Armour, "Hi, I'm Joan and I'm a Workaholic," *USA Today*, May 23, 2007, p. 1B, 2B). Suppose that the survey of 1,564 workaholics included 786 men and 778 women, and the results showed that 707 men and 638 women loved their job because it is stimulating/challenging/provides an adrenaline rush. At the 0.05 level of significance, is the proportion of workaholic men who love their job because it is stimulating/challenging/provides an adrenaline rush different from the proportion of women?

2. Where people turn for news is different for various age groups (extracted from P. Johnson, "Young People Turn to the Web for News," *USA Today*, March 23, 2006, p. 9D). Suppose that a study conducted on this issue was based on 200 respondents who were between the ages of 36 and 50, and 200 respondents who were above age 50. Of the 200 respondents who were between the ages of 36 and 50, 82 got their news primarily from newspapers. Of the 200 respondents who were above age 50, 104 got their news primarily from newspapers. Is there evidence of a significant difference in the proportion that get their news primarily from newspapers between those respondents 36 to 50 years old and those above 50 years old? (Use $\alpha = 0.05$.)

3. Do people of different age groups differ in their response to email messages? A survey by the Center for the Digital Future of the University of Southern California (data extracted from A. Mindlin, "Older E-mail Users Favor Fast Replies," *Drill Down, The New York Times*, July 14, 2008, p. B3) reported that 70.7% of users over 70 years of age believe that email messages should be answered quickly as compared to 53.6% of users 12 to 50 years old. Suppose that the survey was based on 1,000 users over 70 years of age and 1,000 users 12 to 50 years old. At the 0.01 level of significance, is there evidence of a significant difference between the two age groups who believe that email messages should be answered quickly?

4. According to a recent study, when shopping online for luxury goods, men spend a mean of $2,401 as compared to women who spend a mean of $1,527 (extracted from R. A. Smith, "Fashion Online: Retailers Tackle the Gender Gap, *The Wall Street Journal*, March 13, 2008, p. D1, D10). Suppose that the study was based on a sample of 100 men and 100 females and the standard deviation of the amount spent was $1,200 for men and $1,000 for women.
    (a) State the null and alternative hypothesis if you want to determine whether the mean amount spent is different for men and for women.

(b) At the 0.05 level of significance, is there evidence that the mean amount spent is different for men and women?

5. You would like to determine at a level of significance of $\alpha = 0.05$, whether the mean surface hardness of steel intaglio printing plates prepared using a new treatment differs from the mean hardness of plates that are untreated. The following results are from an experiment in which 40 steel plates, 20 treated and 20 untreated, were tested for surface hardness.

Intaglio

Surface Hardness of 20 Untreated Steel Plates and 20 Treated Steel Plates

Untreated		Treated	
164.368	177.135	158.239	150.226
159.018	163.903	138.216	155.620
153.871	167.802	168.006	151.233
165.096	160.818	149.654	158.653
157.184	167.433	145.456	151.204
154.496	163.538	168.178	150.869
160.920	164.525	154.321	161.657
164.917	171.230	162.763	157.016
169.091	174.964	161.020	156.670
175.276	166.311	167.706	147.920

At the 0.05 level of significance, is there evidence of a difference in the mean surface hardness of steel intaglio printing plates that are untreated and those prepared using a new treatment?

6. A problem with a telephone line that prevents a customer from receiving or making calls is upsetting to both the customer and the telephone company. The following data represent samples of 20 problems reported to two different offices of a telephone company and the time to clear these problems (in minutes) from the customers' lines:

Phone

Central Office I Time to Clear Problems (minutes)

1.48	1.75	0.78	2.85	0.52	1.60	4.15	3.97	1.48	3.10
1.02	0.53	0.93	1.60	0.80	1.05	6.32	3.93	5.45	0.97

Central Office II Time to Clear Problems (minutes)

7.55	3.75	0.10	1.10	0.60	0.52	3.30	2.10	0.58	4.02
3.75	0.65	1.92	0.60	1.53	4.23	0.08	1.48	1.65	0.72

Is there evidence of a difference in the mean waiting time between the two offices? (Use α = 0.05.)

7. A newspaper article discussed the opening of a Whole Foods Market in the Time-Warner building in New York City. The following data compared the prices of some kitchen staples at the new Whole Foods Market and at the Fairway supermarket located about 15 blocks from the Time-Warner building:

WholeFoods

Item	Whole Foods	Fairway
Half-gallon milk	2.19	1.35
Dozen eggs	2.39	1.69
Tropicana orange juice (64 oz.)	2.00	2.49
Head of Boston lettuce	1.98	1.29
Ground round, 1 lb.,	4.99	3.69
Bumble Bee tuna, 6 oz. can	1.79	1.33
Granny Smith apples (1 lb.)	1.69	1.49
Box DeCecco linguini	1.99	1.59
Salmon steak, 1 lb.,	7.99	5.99
Whole chicken, per pound	2.19	1.49

Source: Extracted from W. Grimes, "A Pleasure Palace Without the Guilt," *The New York Times*, February 18, 2004, pp. F1, F5.

At the 0.05 level of significance, is there evidence of a difference in the mean price between the Whole Foods Market and the Fairway supermarket?

8. Multiple myeloma, or blood plasma cancer, is characterized by increased blood vessel formulation (angiogenesis) in the bone marrow that is a prognostic factor in survival. One treatment approach used for multiple myeloma is stem cell transplantation with the patient's own stem cells. The following data represent the bone marrow microvessel density for patients who had a complete response to the stem cell transplant, as measured by blood and urine tests. The measurements were taken immediately prior to the stem cell transplant and at the time of the complete response:

Myeloma

Patient	Before	After
1	158	284
2	189	214
3	202	101

Patient	Before	After
4	353	227
5	416	290
6	426	176
7	441	290

Source: Extracted from S. V. Rajkumar, R. Fonseca, T. E. Witzig, M. A. Gertz, and P. R. Greipp, "Bone Marrow Angiogenesis in Patients Achieving Complete Response After Stem Cell Transplantation for Multiple Myeloma," *Leukemia*, 1999, 13, pp. 469–472.

At the 0.05 level of significance, is there evidence of a difference in the mean bone marrow microvessel density before the stem cell transplant and after the stem cell transplant?

Concrete

9. The Concrete file contains data that represent the compressive strength, in thousands of pounds per square inch (psi), of 40 samples of concrete taken two and seven days after pouring.

   Source: Extracted from O. Carrillo-Gamboa and R. F. Gunst, "Measurement-Error-Model Collinearities," *Technometrics*, 34, 1992, pp. 454–464.

   At the 0.01 level of significance, is there evidence that the mean strength is lower at two days than at seven days?

Answers to Test Yourself Problems

1. $Z = 4.5266 > 1.96$ (or p-value $= 0.0000 < 0.05$), reject H_0. There is evidence of a difference between men and women in the proportion of workaholics who love their job because it is stimulating/challenging/provides an adrenaline rush.

2. $Z = -2.2054 < -1.96$ (or p-value $= 0.0274 < 0.05$), reject H_0. There is evidence of a difference in the proportion that get their news primarily from newspapers between those respondents 36 to 50 years old and those above 50 years old.

3. $Z = 7.8837 > 2.58$ (or p-value $= 0.0000 < 0.01$), reject H_0. There is evidence of a difference in the proportion who believe that email messages should be answered quickly between the two age groups.

4. (a) $H_0: \mu_1 = \mu_2$ (The two population means are equal.) The alternative hypothesis is $H_1: \mu_1 \neq \mu_2$ (The two population means are not equal.)
   (b) Because $t = 5.5952 > 1.9720$ (or p-value $= 0.0000 < 0.05$), reject H_0. There is evidence that a difference exists in the mean amount spent between men and women.

5. Because $t = 4.104 > 2.0244$ (or p-value $= 0.0002 < 0.05$), reject H_0. There is evidence that a difference exists in the mean surface hardness of steel intaglio printing plates that are untreated and those prepared using a new treatment.

6. Because $t = 0.3544 < 2.0244$ (or p-value $= 0.7250 > 0.05$), do not reject H_0. There is no evidence that a difference exists in the mean waiting time between the two offices.

7. Because $t = 3.2753 > 2.2622$ (or p-value $= 0.0096 < 0.05$), reject H_0. There is evidence of a difference in the mean price at the Whole Foods Market and the Fairway supermarket.

8. Because $t = 1.8426 < 2.4469$ (or p-value $= 0.1150 > 0.05$), do not reject H_0. There is no evidence of a difference in the mean bone marrow microvessel density before the stem cell transplant and after the stem cell transplant.

9. Because $t = -9.3721 < -2.4258$ (or p-value $= 0.0000 > 0.01$), reject H_0. There is evidence that the mean strength is lower at two days than at seven days.

References

1. Berenson, M. L., D. M. Levine, and T. C. Krehbiel. *Basic Business Statistics: Concepts and Applications*, Eleventh Edition. Upper Saddle River, NJ: Prentice Hall, 2009.

2. Levine, D. M., T. C. Krehbiel, and M. L. Berenson. *Business Statistics: A First Course*, Fifth Edition. Upper Saddle River, NJ: Prentice Hall, 2010.

3. Levine, D. M., D. Stephan, T. C. Krehbiel, and M. L. Berenson. *Statistics for Managers Using Microsoft Excel*, Fifth Edition. Upper Saddle River, NJ: Prentice Hall, 2005.

4. Levine, D. M., P. P. Ramsey, and R. K. Smidt, *Applied Statistics for Engineers and Scientists Using Microsoft Excel and Minitab*. Upper Saddle River, NJ: Prentice Hall, 2001.

5. Microsoft Excel 2007. Redmond, WA: Microsoft Corporation, 2006.

Hypothesis Testing: Chi-Square Tests and the One-Way Analysis of Variance (ANOVA)

9.1 Chi-Square Test for Two-Way Cross-Classification Tables

9.2 One-Way Analysis of Variance (ANOVA): Testing for the Differences Among the Means of More Than Two Groups
Important Equations
One-Minute Summary
Test Yourself

In Chapter 8, you learned several hypothesis tests that you use to analyze differences between two groups. In this chapter, you learn about tests that you can use when you have multiple (two or more) groups.

9.1 Chi-Square Test for Two-Way Cross-Classification Tables

CONCEPT   The hypothesis tests for the difference in the proportion of successes in two or more groups or a relationship between two categorical variables in a two-way cross-classification table.

INTERPRETATION   Recall from Chapter 2, "Presenting Data in Charts and Tables," that a two-way cross-classification table presents the count of joint responses to two categorical variables. The categories of one variable form the rows of the table and the categories of the other variable form the columns. The chi-square test determines whether a relationship exists between the row variable and the column variable.

The null and alternative hypotheses for the two-way cross-classification table are

H_0: (There is no relationship between the row variable and the column variable.)
H_1: (There is a relationship between the row variable and the column variable.)

For the special case of a table that contains only two rows and two columns, the chi-square test becomes equivalent to the Z test for the difference between two proportions discussed in Section 8.1. The null and alternative hypotheses are restated as

$H_0: \pi_1 = \pi_2$ (No difference exists between the two proportions.)

$H_1: \pi_1 \neq \pi_2$ (A difference exists between the two proportions.)

The chi-square test compares the actual count (or frequency) in each cell, the intersection of a row and column, with the frequency that would be expected to occur if the null hypothesis were true. The expected frequency for each cell is calculated by multiplying the row total of that cell by the column total of that cell and dividing by the total sample size:

$$\text{expected frequency} = \frac{(\text{row total})(\text{column total})}{\text{sample size}}$$

Because some differences are positive and some are negative, each difference is squared; then each squared difference is divided by the expected frequency. The results for all cells are then summed to produce a statistic that follows the chi-square distribution.

To use this test, the expected frequency for each cell must be greater than 1.0, except for the special case of a two-way table that has two rows and two columns, in which the expected frequency of each cell should be at least 5. (If, for this special case, the expected frequency is less than 5, you can use alternative tests such as Fisher's exact test [see References 2 and 3].)

WORKED-OUT PROBLEM 1   You want to determine, with a level of significance $\alpha = 0.05$, whether the results of the study concerning the manufacture of silicon chips first presented in Chapter 2 show that a difference exists between the proportions of good and bad wafers that have particles (Particles present = Yes). The data are as follows:

Counts of Particles Found Cross-Classified by Wafer Condition

		Wafer Condition		Total
		Good	Bad	
Particles Present	Yes	14	36	50
	No	320	80	400
	Total	334	116	450

The row 1 total shows that 50 wafers have particles present. The column 1 total shows that 334 wafers are good. The expected frequency for good wafers with particles present is 37.11—the total number of wafers with particles present (50) multiplied by the total number of good wafers (334) and divided by the total, or sample size (450).

$$\text{expected frequency} = \frac{(50)(334)}{450}$$
$$= 37.11$$

The expected frequencies for all four cells are as follows:

		Wafer Condition		
		Good	Bad	Total
Particles Present	Yes	37.11	12.89	50
	No	296.89	103.11	400
	Total	334	116	450

Spreadsheet and calculator test results for this study are as follows:

	A	B	C	D
1	Chi-Square Test			
2				
3	Observed Frequencies			
4		Wafer Condition		
5	Particles Present	Good	Bad	Total
6	Yes	14	36	50
7	No	320	80	400
8	Total	334	116	450
9				
10	Expected Frequencies			
11		Wafer Condition		
12	Particles Present	Good	Bad	Total
13	Yes	37.11111	12.88889	50
14	No	296.8889	103.1111	400
15	Total	334	116	450
16				
17	Data			
18	Level of Significance	0.05		
19	Number of Rows	2		
20	Number of Columns	2		
21	Degrees of Freedom	1		
22				
23	Results			
24	Critical Value	3.8415		
25	Chi-Square Test Statistic	62.8123		
26	p-Value	0.0000		
27	Reject the null hypothesis			
28				
29	Expected frequency assumption is met.			

χ^2-Test
$\chi^2 = 62.81230642$
$P = 2.273737\text{E} - 15$
$df = 1$

The results show that the p-value for this chi-square test is 0.0000 (2.273737E-15 is an extremely small number that you can consider as the value zero). Because the p-value is less than the level of significance α, 0.05,

you reject the null hypothesis. You conclude that a relationship exists
between having a particle on the wafer and the condition of the wafer. You
assert that a significant difference exists between good and bad wafers in the
proportion of wafers that have particles.

WORKED-OUT PROBLEM 2   You decide to use the critical value
approach for the study concerning the manufacture of silicon chips. The
computed chi-square statistic is 62.81 (see the results above). The number of
degrees of freedom for the chi-square test equals the number of rows minus 1
multiplied by the number of columns minus 1:

Degrees of freedom = (Number of rows – 1) × (Number of columns – 1)

Using the table of the chi-square distribution (Table C.3), with $\alpha = 0.05$ and
the degrees of freedom = (2 – 1) (2 – 1) = 1, the critical value of chi-square is
equal to 3.841. Because 62.81 > 3.841, you reject the null hypothesis.

calculator keys

Chi-Square Tests

To enter a table of observed frequencies:

Press [**2nd**][**x⁻¹**][◀] and press [**ENTER**] to display the
MATRIX[A] screen.

- Type the number of rows, press [▶], type the number of
  columns, and press [▶].

- In the table-like entry area that appears, enter the
  observed frequency of each cell. (Press [**ENTER**] after
  each entry.)

When your data entry is completed, press [**2nd**][**MODE**].
Your table of observed frequencies is stored in the matrix
variable [A].

To perform a chi-square test:

Press [**STAT**][▶] (to display the **TESTS** menu) and select
C: χ^2-**Test** and press [**ENTER**] to display the χ^2-**Test** screen.
Verify that **Observed** is [A] and that **Expected** is [B]. Select
Calculate and press [**ENTER**].

spreadsheet solution

Chi-Square Tests

Chapter 9 Chi-Square performs a chi-square test based on a table of observed frequencies and a level of significance that you enter. Choose either the **ChiSquare 2 x 2**, **ChiSquare 2 x 3**, **ChiSquare 3 x 4**, or **ChiSquare 5 x 3** worksheets, as appropriate, and then fill in the observed frequencies table in the worksheet you chose.

Spreadsheet Tips FT10 and FT11 in Appendix D explain the spreadsheet functions used to calculate the critical value and the *p*-value.

WORKED-OUT PROBLEM 3   Fast-food chains are evaluated on many variables and the results are summarized periodically in *QSR Magazine*. One important variable is the accuracy of the order. You seek to determine, with a level of significance of $\alpha = 0.05$, whether a difference exists in the proportions of food orders filled correctly at Burger King, Wendy's, and McDonald's. You use the following data that report the results of placing an order consisting of a main item, a side item, and a drink.

		Fast Food Chain		
		Burger King	**Wendy's**	**McDonald's**
Order Filled Correctly	Yes	440	430	422
	No	60	70	78
	Total	500	500	500

The null and alternative hypotheses are as follows:

H_0: $\pi_1 = \pi_2 = \pi_3$ (No difference exists in the proportion of correct orders among Burger King, Wendy's, and McDonald's.)

H_1: $\pi_1 \neq \pi_2 \neq \pi_3$ (A difference exists in the proportion of correct orders among Burger King, Wendy's, and McDonald's.)

Spreadsheet and calculator results for this study are as follows:

	A	B	C	D	E
1	Chi-Square for 2 X 3				
2					
3		Observed Frequencies			
4		Fast Food Chain			
5	Order Filled Correctly	Burger King	Wendy's	McDonald's	Total
6	Yes	440	430	422	1292
7	No	60	70	78	208
8	Total	500	500	500	1500
9					
10		Expected Frequencies			
11		Fast Food Chain			
12	Order Filled Correctly	Burger King	Wendy's	McDonald's	Total
13	Yes	430.6667	430.6667	430.6667	1292
14	No	69.3333	69.3333	69.3333	208
15	Total	500	500	500	1500
16					
17	Data				
18	Level of Significance	0.05			
19	Number of Rows	2			
20	Number of Columns	3			
21	Degrees of Freedom	2			
22					
23	Results				
24	Critical Value	5.9915			
25	Chi-Square Test Statistic	2.7239			
26	p-Value	0.2562			
27	Do not reject the null hypothesis				
28					
29	Expected frequency assumption				
30	is met.				

```
X²-Test
 X²=2.723862824
 P=.2561655376
 df=2
```

Because the p-value for this chi-square test, 0.2562, is greater than the level of significance α of 0.05, you cannot reject the null hypothesis. Insufficient evidence exists of a difference in the proportion of correct orders filled among Burger King, Wendy's, and McDonald's.

WORKED-OUT PROBLEM 4   Using the critical value approach for the same problem, the computed chi-square statistic is 2.72 (see the results). At the 0.05 level of significance with the 2 degrees of freedom [(2 – 1) (3 – 1) = 2], the chi-square critical value from Table C.3 is 5.991. Because the computed test statistic is less than 5.991, you cannot reject the null hypothesis.

WORKED-OUT PROBLEM 5   You want to determine (at the 0.05 level of significance) whether the media that people turn to for news is different for various age groups. A study indicated where different age groups primarily get their news:

		Age Group		
		Under 36	36–50	50+
Media	**Local TV**	107	119	133
	National TV	73	102	127
	Radio	75	97	109
	Local newspaper	52	79	107
	Internet	95	83	76

The null and alternative hypotheses are

H_0: No relationship exists between the age group and the media where people get their news.

H_1: A relationship exists between the age group and the media where people get their news.

Spreadsheet and calculator results are as follows:

	A	B	C	D	E
1	Chi-Square Test				
2					
3		Observed Frequencies			
4			AGE GROUP		
5	MEDIA	Under 36	36-50	50+	Total
6	Local TV	107	119	133	359
7	National TV	73	102	127	302
8	Radio	75	97	109	281
9	Local newspaper	52	79	107	238
10	Internet	95	83	76	254
11	Total	402	480	552	1434
12					
13		Expected Frequencies			
14			AGE GROUP		
15	MEDIA	Under 36	36-50	50+	Total
16	Local TV	100.6402	120.1674	138.1925	359
17	National TV	84.6611	101.0879	116.2510	302
18	Radio	78.7741	94.0586	108.1674	281
19	Local newspaper	66.7197	79.6653	91.6151	238
20	Internet	71.2050	85.0209	97.7741	254
21	Total	402	480	552	1434
22					
23	Data				
24	Level of Significance	0.05			
25	Number of Rows	5			
26	Number of Columns	3			
27	Degrees of Freedom	8			
28					
29	Results				
30	Critical Value	15.5073			
31	Chi-Square Test Statistic	22.1812			
32	p-Value	0.0046			
33	Reject the null hypothesis				
34					
35	Expected frequency assumption is met.				

```
X²-Test
 X²=22.18121667
 P=.0045909966
 df=8
```

Because the p-value for this chi-square test, 0.0046, is less than the level of significance α of 0.05, you reject the null hypothesis. Evidence exists of a relationship between the age group and the media where people get their news. It appears that people 50 and above are more likely to get their news from the newspaper or TV, while people under 36 are more likely to get their news from the Internet.

WORKED-OUT PROBLEM 6   Using the critical value approach for the same problem, the computed chi-square statistic is 22.1812 (see the previous results). At the 0.05 level of significance with the 8 degrees of freedom [(5 – 1) (3 – 1) = 8], the chi-square critical value from Table C.3 is 15.5073. Because the computed test statistic of 22.1812 is greater than 15.5073, you reject the null hypothesis.

You need the subscripted symbols for the **observed cell frequencies**, f_0, and the **expected cell frequencies**, f_e, to write the equation for the chi-square test for a two-way cross classification table:

$$\chi^2 = \sum_{all\ cells} \frac{(f_0 - f_e)^2}{f_e}$$

For the study concerning the manufacture of silicon chips (WORKED-OUT PROBLEM 1), the calculations are as follows:

f_o	f_e	$(f_o - f_e)$	$(f_o - f_e)^2$	$(f_o - f_e)^2/f_e$
14	37.11	-23.11	534.0721	14.3916
320	296.89	23.11	534.0721	1.7989
36	12.89	23.11	534.0721	41.4331
80	103.11	-23.11	534.0721	5.1796
				62.8032

Using the level of significance $\alpha = 0.05$, with $(2-1)(2-1) = 1$ degree of freedom, from Table C.3, the critical value is 3.841. Because $62.80 > 3.841$, you reject the null hypothesis.

9.2 One-Way Analysis of Variance (ANOVA): Testing for the Differences Among the Means of More Than Two Groups

Many analyses involve experiments in which you want to test whether differences exist in the means of more than two groups. Evaluating differences between groups is often viewed as a one-factor experiment (also known as a **completely randomized design**) in which the variable that defines the groups is called the factor of interest. A factor of interest can have several *numerical levels* such as baking temperature (e.g., 300°, 350°, 400°, 450°) for an industrial process study, or a factor can have several *categorical levels* such as type of learning materials (Type A, Type B, Type C) for an educational research study.

One-Way ANOVA

CONCEPT The hypothesis test that simultaneously compares the differences among the population means of more than two groups in a one-factor experiment.

INTERPRETATION Unlike the t test, which compares differences in two means, the analysis of variance simultaneously compares the differences among the means of more than two groups. Although ANOVA is an acronym for **AN**alysis **O**f **VA**riance, the term is misleading, because the objective in the Analysis of Variance is to analyze differences among the group means, *not* the variances. The null and alternative hypotheses are

H_0: (All the population means are equal.)

H_1: (Not all the population means are equal.)

In ANOVA, the total variation in the values is subdivided into variation that is due to differences among the groups and variation that is due to variation within the groups (see the following figure). Within group variation is called **experimental error**, and the variation between the groups that represents variation due to the factor of interest is called the **treatment effect**.

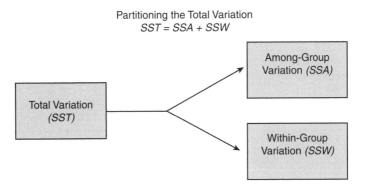

Partitioning the Total Variation
$SST = SSA + SSW$

The **sum of squares total** (*SST*) is the total variation that represents the sum of the squared differences between each individual value and the mean of all the values:

SST = Sum of (Each value − Mean of all values)^2

The **sum of squares among groups** (*SSA*) is the among-group variation that represents the sum of the squared differences between the sample mean of each group and the mean of all the values, weighted by the sample size in each group:

SSA = sum of [(Sample size in each group) (Group mean − Mean of all values)^2]

The **sum of squares within groups** (*SSW*) is the within-group variation that measures the difference between each value and the mean of its own group and sums the squares of these differences over all groups:

$$SSW = \text{Sum of } [(\text{Each value in the group} - \text{Group mean})^2]$$

The Three Variances of ANOVA

ANOVA derives its name from the fact that the differences between the means of the groups are analyzed by comparing variances. In Section 3.3 on page 55, the variance was calculated as a sum of squared differences around the mean divided by the sample size minus 1.

$$\text{Variance} = S^2 = \frac{\text{Sum of squared differences around the mean}}{\text{sample size - 1}}$$

This sample size minus 1 represents the actual number of values that are free to vary once the mean is known and is called the **degrees of freedom**.

In the analysis of variance, there are three different variances: the variance among groups, the variance within groups, and the total variance. These variances are referred to in the analysis-of-variance terminology as **mean squares**. The **mean square among groups** (*MSA*) is equal to the sum of squares among groups (*SSA*) divided by the number of groups minus 1. The **mean square within groups** (*MSW*) is equal to the sum of squares within groups (*SSW*) divided by the sample size minus the number of groups. The **mean square total** (*MST*) is equal to the sum of squares total (*SST*) divided by the sample size minus 1.

To test the null hypothesis

H_0: All the population means are equal.

against the alternative

H_1: Not all the population means are equal.

you calculate the test statistic *F*, which follows the *F* distribution (see Table C.4), as the ratio of two of the variances, *MSA* to *MSW*.

$$F = \frac{MSA}{MSW}$$

ANOVA Summary Table

The results of an analysis of variance are usually displayed in an ANOVA summary table. The entries in this table include the sources of variation (among-group, within-group, and total), the degrees of freedom, the sums of squares, the mean squares (or variances), and the F test statistic. A p-value is often included in the table when software is used.

Analysis of Variance Summary Table

Source	Degrees of Freedom	Sum of Squares	Mean Square (Variance)	F
Among groups	number of groups − 1	SSA	$MSA = \dfrac{SSA}{\text{number of groups - 1}}$	$F = \dfrac{MSA}{MSW}$
Within groups	sample size − number of groups	SSW	$MSW = \dfrac{SSW}{\text{sample size - number of groups}}$	
Total	sample size − 1	SST		

important point

After performing a one-way ANOVA and finding a significant difference among groups, you do not know which groups are significantly different. All that is known is that sufficient evidence exists to state that the population means are not all the same. To determine exactly which groups differ, all possible pairs of groups need to be compared. Many statistical procedures for making these comparisons have been developed (see References 1, 5, 6, 8).

WORKED-OUT PROBLEM 7   You want to determine, with a level of significance $\alpha = 0.05$, whether differences exist among three sets of mathematics learning materials (labeled A, B, and C). You devise an experiment that randomly assigns 24 students to one of the three sets of materials. At the end of a school year, all 24 students are given the same standardized mathematics test that is scored on a 0 to 100 scale. The results of that test are as follows:

Math

A	B	C
87	58	81
80	63	62
74	64	70
82	75	64
74	70	70
81	73	72
97	80	92
71	62	63

Spreadsheet and calculator results for these data are as follows:

	A	B	C	D	E	F	G
1	**Anova: Single Factor for Learning Materials Study**						
2							
3	**SUMMARY**						
4	*Groups*	*Count*	*Sum*	*Average*	*Variance*		
5	A	8	646	80.75	70.2143		
6	B	8	545	68.125	56.9821		
7	C	8	574	71.75	104.7857		
8							
9							
10	**ANOVA**						
11	*Source of Variation*	*SS*	*df*	*MS*	*F*	*P-value*	*F crit*
12	Between Groups	676.0833	2	338.0417	4.371565	0.0259	3.4668
13	Within Groups	1623.8750	21	77.3274			
14							
15	Total	2299.9583	23				

```
One-way ANOVA              One-way ANOVA
 F=4.37156493              ↑ MS=338.041667
 P=.0258667931               Error
 Factor                       df=21
  df=2                        SS=1623.875
  SS=676.083333               MS=77.327381
↓ MS=338.041667              SxP=8.79359886
```

Because the p-value for this test, 0.0259, is less than the level of significance $\alpha = 0.05$, you reject the null hypothesis. You can conclude that the mean scores are not the same for all the sets of mathematics materials. From the results, you see that the mean for materials A is 80.75, for materials B it is 68.125, and for materials C it is 71.75. It appears that the mean score is higher for materials A than for materials B and C.

WORKED-OUT PROBLEM 8 Using the critical value approach for the same problem, the computed F statistic is 4.37 (see the earlier results). To determine the critical value of F, you refer to the table of the F statistic (Table C.4). This table requires these degrees of freedom:

- The numerator degrees of freedom, equal to the number of groups minus 1

- The denominator degrees of freedom, equal to the sample size minus the number of groups

With three groups and a sample size of 24, the numerator degrees of freedom are 2 (3 – 1 = 2), and the denominator degrees of freedom are 21 (24 – 3 = 21). With the level of significance $\alpha = 0.05$, the critical value of F from Table C.4 is 3.47 (also shown in the spreadsheet results). Because the decision rule is to reject H_0 if $F >$ critical value of F, and $F = 4.37 > 3.47$, you reject the null hypothesis.

equation
blackboard
(optional)

interested
in
math?

To form the equations for the three mean squares and the test statistic F, you assemble these symbols:

- $\bar{\bar{X}}$, pronounced as "X double bar," that represents the overall or grand mean
- a subscripted X Bar, $\bar{X}_j$, that represents the mean of a group
- a double-subscripted uppercase italic X, X_{ij}, that represents individual values in group j
- a subscripted lowercase italic n, n_j, that represents the sample size in a group
- a lowercase italic n, n, that represents the total sample size (sum of the sample sizes of each group)
- a lowercase italic c, c, that represents the number of groups

First, you form the equation for the grand mean as

$$\bar{\bar{X}} = \frac{\sum\limits_{j=1}^{c}\sum\limits_{i=1}^{n_j} X_{ij}}{n} = \text{grand mean}$$

$\bar{X}_j$ = sample mean of group j

X_{ij} = ith value in group j

n_j = number of values in group j

n = total number of values in all groups combined

(that is, $n = n_1 + n_2 + \cdots + n_c$)

c = number of groups of the factor of interest

With $\bar{\bar{X}}$ defined, you then form the equations that define the sum of squares total, SST, the sum of squares among groups, SSA, and the sum of squares within groups, SSW:

$$SST = \sum_{j=1}^{c}\sum_{i=1}^{n_j}(X_{ij} - \bar{\bar{X}})^2$$

$$SSA = \sum_{j=1}^{c} n_j(\bar{X}_j - \bar{\bar{X}})^2$$

$$SSW = \sum_{j=1}^{c}\sum_{i=1}^{n_j}(X_{ij} - \bar{X}_j)^2$$

(continues)

Next, using these definitions, you form the equations for the mean squares:

$$MSA = \frac{SSA}{\text{number of groups - 1}}$$

$$MSW = \frac{SSW}{\text{sample size - number of groups}}$$

$$MST = \frac{SST}{\text{sample size - 1}}$$

Finally, using the definitions of MSA and MSW, you form the equation for the test statistic F:

$$F = \frac{MSA}{MSW}$$

Using the data from WORKED-OUT EXAMPLE 7,

$$\overline{\overline{X}} = \frac{1,765}{24} = 73.5417$$

$$SST = \sum_{j=1}^{c} \sum_{i=1}^{n_j} (X_{ij} - \overline{\overline{X}})^2 = (87 - 73.5417)^2 + \cdots + (71 - 73.5417)^2$$

$$+ (58 - 73.5417)^2 + \cdots + (62 - 73.5417)^2$$

$$+ (81 - 73.5417)^2 + \cdots + (63 - 73.5417)^2 = 2,299.9583$$

$$SSA = \sum_{j=1}^{c} n_j (\overline{X}_j - \overline{\overline{X}})^2 = 8(80.75 - 73.5417)^2$$

$$+ 8(68.125 - 73.5417)^2 + 8(71.75 - 73.5417)^2 = 676.0833$$

$$SSW = \sum_{j=1}^{c} \sum_{i=1}^{n_j} (X_{ij} - \overline{X}_j)^2 = (87 - 80.75)^2 + \cdots$$

$$+ (71 - 80.75)^2 + (58 - 68.125)^2 + \cdots + (62 - 68.125)^2$$

$$+ (81 - 71.75)^2 + \cdots + (63 - 71.75)^2 = 1,623.875$$

Then,

$$MSA = \frac{SSA}{\text{number of groups - 1}} = \frac{676.0833}{2} = 338.0417$$

$$MSW = \frac{SSW}{\text{sample size - number of groups}} = \frac{1{,}623.8750}{21} = 77.3274$$

Then,

$$F = \frac{MSA}{MSW} = \frac{338.0417}{77.3274} = 4.3716$$

Because $F = 4.3716 > 3.47$, you reject H_0.

calculator keys

One-Way ANOVA

Enter the data values of each group into separate list variables
(see Chapter 1). Then, press [STAT][◄] (to display the
TESTS menu) and select H:ANOVA(and press [ENTER] to
display the ANOVA function. Enter the names of the list variables you used, separated by commas, and press [ENTER].
ANOVA results appear over two screens. Press [▼] and [▲]
to scroll through the results.

(Note: In TI-83 series calculators, the ANOVA choice will be
labeled as F:ANOVA(.)

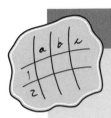

spreadsheet solution

One-Way ANOVA

Chapter 9 One-Way ANOVA ATP (shown on page 190) contains the results of using the Analysis ToolPak **ANOVA: Single
Factor** procedure to perform a one-way ANOVA of the learning materials study data of WORKED-OUT PROBLEMS 7 and
8.

Spreadsheet Tip ATT5 in Appendix D explains how to use the
ANOVA: Single Factor procedure.

WORKED-OUT PROBLEM 9   A pet food company is looking to expand its product line beyond its current kidney and shrimp-based cat foods. The company developed two new products, one based on chicken livers, and the other based on salmon. The company conducted an experiment to compare the two new products with its two existing ones as well as a generic beef-based product sold in a supermarket chain.

For the experiment, a sample of 50 cats from the population at a local animal shelter was selected. Ten cats were randomly assigned to each of the five products being tested. Each of the cats was then presented with three ounces of the selected food in a dish at feeding time. The researchers defined the variable to be measured as the number of ounces of food that the cat consumed within a 10-minute time interval that began when the filled dish was presented. The results for this experiment are summarized in the following table.

CatFood

Kidney	Shrimp	Chicken Liver	Salmon	Beef
2.37	2.26	2.29	1.79	2.09
2.62	2.69	2.23	2.33	1.87
2.31	2.25	2.41	1.96	1.67
2.47	2.45	2.68	2.05	1.64
2.59	2.34	2.25	2.26	2.16
2.62	2.37	2.17	2.24	1.75
2.34	2.22	2.37	1.96	1.18
2.47	2.56	2.26	1.58	1.92
2.45	2.36	2.45	2.18	1.32
2.32	2.59	2.57	1.93	1.94

Spreadsheet and calculator results for this study are as follows:

	A	B	C	D	E	F	G
1	Anova: Single Factor for Cat Food Study						
2							
3	SUMMARY						
4	*Groups*	*Count*	*Sum*	*Average*	*Variance*		
5	Kidney	10	24.56	2.456	0.0148		
6	Shrimp	10	24.09	2.409	0.0253		
7	Chicken Liver	10	23.68	2.368	0.0263		
8	Salmon	10	20.28	2.028	0.0544		
9	Beef	10	17.54	1.754	0.0990		
10							
11							
12	ANOVA						
13	Source of Variation	*SS*	*df*	*MS*	*F*	*P-value*	*F crit*
14	Between Groups	3.6590	4	0.9147	20.8054	9.15E-10	2.5787
15	Within Groups	1.9785	45	0.0440			
16							
17	Total	5.6375	49				

```
One-way ANOVA          One-way ANOVA
 F=20.80541221          ↑ MS=.91474
 P=9.148998E-10          Error
 Factor                   df=45
  df=4                    SS=1.97849
  SS=3.65896              MS=.043966444
↓ MS=.91474             SxP=.209681769
```

Because the *p*-value for this test, reported as 9.15E-10, a very small value that you can consider to be zero, is less than the level of significance $\alpha = 0.05$, you reject the null hypothesis. You conclude that evidence of a difference exists in the mean amount of food eaten among the five types of cat foods.

WORKED-OUT PROBLEM 10 Using the critical value approach for the same problem, the computed *F* statistic is 20.8054. At the level of significance $\alpha = 0.05$, with 4 degrees of freedom in the numerator $(5 - 1)$ and 45 degrees of freedom in the denominator $(50 - 5)$, the critical value of *F* from the Excel results is 2.5787. Because the computed *F* test statistic 20.8054 is greater than 2.5787, you reject the null hypothesis.

One-Way ANOVA Assumptions

There are three major assumptions you must make to use the one-way ANOVA *F* test: randomness and independence, normality, and homogeneity of variance.

The first assumption, randomness and independence, always must be met, because the validity of your experiment depends on the random sampling or random assignment of items or subjects to groups. Departures from this assumption can seriously affect inferences from the analysis of variance. These problems are discussed more thoroughly in References 7 and 8.

The second assumption, normality, states that the values in each group are selected from normally distributed populations. The one-way ANOVA *F* test is not very sensitive to departures from this assumption of normality. As long as the distributions are not very skewed, the level of significance of the ANOVA *F* test is usually not greatly affected by lack of normality, particularly for large samples. When only the normality assumption is seriously violated, nonparametric alternatives to the one-way ANOVA *F* test are available (see References 1, 5, and 6).

The third assumption, equality of variances, states that the variance within each population should be equal for all populations. Although the one-way ANOVA *F* test is relatively robust or insensitive with respect to the assumption of equal group variances, large departures from this assumption can seriously affect the level of significance and the power of the test. Therefore, various procedures have been developed to test the assumption of homogeneity of variance (see References 1, 4, 5, and 6).

One way to evaluate the assumptions is to construct a side-by-side box-and-whisker plot of the groups to study their central tendency, variation, and shape.

Other Experimental Designs

The one-way analysis of variance is the simplest type of experimental design, because it considers only one factor of interest. More complicated experimental designs examine at least two factors of interest simultaneously. For more information on these designs, see References 1, 4, 7, and 8.

Important Equations

Chi-square test for a two-way cross-classification table:

$$(9.1) \quad \chi^2 = \sum_{all\ cells} \frac{(f_0 - f_e)^2}{f_e}$$

ANOVA calculations

$$(9.2) \quad SST = \sum_{j=1}^{c} \sum_{i=1}^{n_j} (X_{ij} - \bar{\bar{X}})^2$$

$$(9.3) \quad SSA = \sum_{j=1}^{c} n_j (\bar{X}_j - \bar{\bar{X}})^2$$

$$(9.4) \quad SSW = \sum_{j=1}^{c} \sum_{i=1}^{n_j} (X_{ij} - \bar{X}_j)^2$$

$$(9.5) \quad MSA = \frac{SSA}{number\ of\ groups - 1}$$

$$(9.6) \quad MSW = \frac{SSW}{\text{sample size - number of groups}}$$

$$(9.7) \quad MST = \frac{SST}{\text{sample size - 1}}$$

$$(9.8) \quad F = \frac{MSA}{MSW}$$

One-Minute Summary

Tests for the differences among more than two groups:

- If your data are categorical, use chi-square (χ^2) tests (can also use for two groups).
- If your data are numerical and if you have one factor, use the one-way ANOVA.

Test Yourself

Short Answers

1. In a one-way ANOVA, if the F test statistic is greater than the critical F value, you:
   (a) reject H_0 because there is evidence all the means differ
   (b) reject H_0 because there is evidence at least one of the means differs from the others
   (c) do not reject H_0 because there is no evidence of a difference in the means
   (d) do not reject H_0 because one mean is different from the others

2. In a one-way ANOVA, if the p-value is greater than the level of significance, you:
   (a) reject H_0 because there is evidence all the means differ
   (b) reject H_0 because there is evidence at least one of the means differs from the others.
   (c) do not reject H_0 because there is no evidence of a difference in the means
   (d) do not reject H_0 because one mean is different from the others

3. The F test statistic in a one-way ANOVA is:
   (a) MSW/MSA
   (b) SSW/SSA
   (c) MSA/MSW
   (d) SSA/SSW

4. In a one-way ANOVA, the null hypothesis is always:
   (a) all the population means are different
   (b) some of the population means are different
   (c) some of the population means are the same
   (d) all of the population means are the same

5. A car rental company wants to select a computer software package for its reservation system. Three software packages (A, B, and C) are commercially available. The car rental company will choose the package that has the lowest mean number of renters for whom a car is not available at the time of pickup. An experiment is set up in which each package is used to make reservations for five randomly selected weeks. How should the data be analyzed?
   (a) Chi-square test for differences in proportions
   (b) One-way ANOVA F test
   (c) t test for the differences in means
   (d) t test for the mean difference

The following should be used to answer Questions 6 through 9:

For fast-food restaurants, the drive-through window is an increasing source of revenue. The chain that offers that fastest service is considered most likely to attract additional customers. In a study of 20 drive-through times (from menu board to departure) at 5 fast-food chains, the following ANOVA table was developed.

Source	DF	Sum of Squares	Mean Squares	F
Among Groups (Chains)		6,536	1,634.0	12.51
Within Groups (Chains)	95		130.6	
Total	99	18,943		

6. Referring to the preceding table, the Among Groups degrees of freedom is:
   (a) 3
   (b) 4
   (c) 12
   (d) 16

7. Referring to the preceding table, the within groups sum of squares is:
   (a) 12,407
   (b) 95
   (c) 130.6
   (d) 4

8. Referring to the preceding table, the within groups mean squares is:
   (a) 12,407
   (b) 95
   (c) 130.6
   (d) 4

9. Referring to the preceding table, at the 0.05 level of significance, you:
   (a) do not reject the null hypothesis and conclude that no difference exists in the mean drive-up time between the fast-food chains
   (b) do not reject the null hypothesis and conclude that a difference exists in the mean drive-up time between the fast-food chains
   (c) reject the null hypothesis and conclude that a difference exists in the mean drive-up time between the fast-food chains
   (d) reject the null hypothesis and conclude that no difference exists in the mean drive-up time between the fast-food chains

10. When testing for independence in a contingency table with three rows and four columns, there are _____ degrees of freedom.
    (a) 5
    (b) 6
    (c) 7
    (d) 12

11. In testing a hypothesis using the chi-square test, the theoretical frequencies are based on the:
    (a) null hypothesis
    (b) alternative hypothesis
    (c) normal distribution
    (d) t distribution

12. An agronomist is studying three different varieties of tomato to determine whether a difference exists in the proportion of seeds that germinate. Random samples of 100 seeds of each of three varieties are subjected to the same starting conditions. How should the data be analyzed?
    (a) Chi-square test for differences in proportions
    (b) One-way ANOVA F test
    (c) t test for the differences in means
    (d) t test for the mean difference

Answer True or False:

13. A test for the difference between two proportions can be performed using the chi-square distribution.

14. The analysis-of-variance (ANOVA) tests hypotheses about the difference between population proportions.

15. The one-way analysis-of-variance (ANOVA) tests hypotheses about the difference between population means.

16. The one-way analysis-of-variance (ANOVA) tests hypotheses about the difference between population variances.

17. The Mean Squares in an ANOVA can never be negative.

18. In a one-factor ANOVA, the Among sum of squares and Within sum of squares must add up to the total sum of squares.

19. If you use the chi-square method of analysis to test for the difference between two proportions, you must assume that there are at least five observed frequencies in each cell of the contingency table.

20. If you use the chi-square method of analysis to test for independence in a contingency table with more than two rows and more than two columns, you must assume that there is at least one theoretical frequency in each cell of the contingency table.

Answers to Test Yourself Short Answers

1. b	11. a
2. c	12. a
3. c	13. True
4. d	14. False
5. b	15. True
6. b	16. False
7. a	17. True
8. c	18. True
9. c	19. True
10. b	20. True

Problems

1. Technology has led to the rise of extreme workers who are on the job 60 hours or more per week. One of the reasons cited by employees as to why they worked long hours was that they loved their job because it is stimulating/challenging/provides an adrenaline rush (extracted from S. Armour, "Hi, I'm Joan and I'm a Workaholic," *USA Today*, May 23, 2007, p. 1B, 2B). Suppose that the survey of 1,564 workaholics

included 786 men and 778 women, and the results showed that 707 men and 638 women loved their job because it is stimulating/challenging/provides an adrenaline rush.

   (a) At the 0.05 level of significance, is the proportion of workaholic men who love their job because it is stimulating/challenging/provides an adrenaline rush different from the proportion of women?

   (b) Compare the results in (a) to those of Problem 1 in Chapter 8 on page 177.

2. Do people of different age groups differ in their response to email messages? A survey by the Center for the Digital Future of the University of Southern California reported that 70.7% of users over 70 years of age believe that email messages should be answered quickly as compared to 53.6% of users 12 to 50 years old. Suppose that the survey is based on 1,000 users over 70 years of age and 1,000 users 12 to 50 years old.

Data extracted from A. Mindlin, "Older E-mail Users Favor Fast Replies," *Drill Down*, *The New York Times*, July 14, 2008, p. B3.

   (a) At the 0.01 level of significance, is there evidence of a significant difference between the two age groups who believe that email messages should be answered quickly?

   (b) Compare the results in (a) to those of Problem 3 in Chapter 8 on page 177.

3. The health-care industry and consumer advocacy groups are at odds over the sharing of a patient's medical records without the patient's consent. The health-care industry believes that no consent should be necessary to openly share data among doctors, hospitals, pharmacies, and insurance companies. Suppose a study is conducted in which 600 patients are randomly assigned, 200 each, to three "organizational groupings"—insurance companies, pharmacies, and medical researchers. Each patient is given material to read about the advantages and disadvantages concerning the sharing of medical records within the assigned "organizational grouping." Each patient is then asked "would you object to the sharing of your medical records with ..." and the results are recorded in the following cross-classification table.

Object to Sharing Information	Organizational Grouping		
	Insurance	Pharmacies	Research
Yes	40	80	90
No	160	120	110

At the 0.05 level of significance, is there a difference in the objection to sharing information among the organizational groupings?

4. You seek to determine, with a level of significance of $\alpha = 0.05$, whether there was a relationship between numbers selected for the Vietnam War

era military draft lottery system and the time of the year a man was born. The following shows how many low (1–122), medium (123–244), and high (245–366) numbers were drawn for birth dates in each quarter of the year.

		Quarter of Year				
		Jan–Mar	Apr–Jun	Jul–Sep	Oct–Dec	Total
	Low	21	28	35	38	122
Number Set	**Medium**	34	22	29	37	122
	High	36	41	28	17	122
	Total	91	91	92	92	366

At the 0.05 level of significance, is there a relationship between draft number and quarter of the year?

5. You want to determine, with a level of significance $\alpha = 0.05$, whether differences exist among the four plants that fill boxes of a particular brand of cereal. You select samples of 20 cereal boxes from each of the four plants. The weights of these cereal boxes (in grams) are as follows.

Cereals

Plant 1		Plant 2		Plant 3		Plant 4	
361.43	364.78	370.26	360.27	367.53	390.12	361.95	369.36
368.91	376.75	357.19	362.54	388.36	335.27	381.95	363.11
365.78	353.37	360.64	352.22	359.33	366.37	383.90	400.18
389.70	372.73	398.68	347.28	367.60	371.49	358.07	358.61
390.96	363.91	380.86	350.43	358.06	358.01	382.40	370.87
372.62	375.68	334.95	376.50	369.93	373.18	386.20	380.56
390.69	380.98	359.26	369.27	355.84	377.40	373.47	376.21
364.93	354.61	389.56	377.36	382.08	396.30	381.16	380.97
387.13	378.03	371.38	368.50	381.45	354.82	379.41	365.78
360.77	374.24	373.06	363.86	356.20	383.78	382.01	395.55

What conclusions can you reach?

6. Integrated circuits are manufactured on silicon wafers through a process that involves a series of steps. An experiment was carried out to study the effect on the yield of using three methods in the cleansing step. The results (coded to maintain confidentiality) are as follows:

Yield

New1	New2	Standard
38	29	31
34	35	23
38	34	38
34	20	29
19	35	32
28	37	30

Source: Extracted from J. Ramirez and W. Taam, "An Autologistic Model for Integrated Circuit Manufacturing," *Journal of Quality Technology*, 2000, 32, pp. 254–262.

At the 0.05 level of significance, is there a difference in the yield among the methods used in the cleansing steps?

7. A sporting goods manufacturing company wanted to compare the distance traveled by golf balls produced using each of four different designs. Ten balls were manufactured with each design and each ball was tested at a testing facility using a robot to hit the balls. The results (distance traveled in yards) for the four designs were as follows:

GolfBall

Design1	Design2	Design3	Design4
206.32	217.08	226.77	230.55
207.94	221.43	224.79	227.95
206.19	218.04	229.75	231.84
204.45	224.13	228.51	224.87
209.65	211.82	221.44	229.49
203.81	213.90	223.85	231.10
206.75	221.28	223.97	221.53
205.68	229.43	234.30	235.45
204.49	213.54	219.50	228.35
210.86	214.51	233.00	225.09

At the 0.05 level of significance, is there evidence of a difference in the mean distances traveled by the golf balls with different designs?

Answers to Test Yourself Problems

1. (a) Because the p-value for this chi-square test, 0.0000, is less than the level of significance α of 0.05 (or the chi-square statistic = 20.4903 > 3.841), you reject the null hypothesis. There is evidence of a difference between men and women in the proportion of workaholics who love their job because it is stimulating/challenging/provides an adrenaline rush.

   (b) The results are the same because the chi-square statistic with one degree of freedom is the square of the Z statistic.

2. Because the p-value for this chi-square test, 0.0000, is less than the level of significance α of 0.01 (or the chi-square statistic = 62.152 > 6.635), you reject the null hypothesis. There is evidence of a difference in the proportion who believe that email messages should be answered quickly between the two age groups.

   (b) The results are the same because the chi-square statistic with one degree of freedom is the square of the Z statistic.

3. Because the p-value for this chi-square test, 0.0000, is less than the level of significance α of 0.05 (or the chi-square statistic = 30.7692 > 5.991), you reject the null hypothesis. There is evidence of a difference in the objection to sharing information among the organizational groupings.

4. Because the p-value for this chi-square test, 0.0021, is less than the level of significance α of 0.05 (or the chi-square statistic = 20.6804 > 12.5916), you reject the null hypothesis. Evidence exists of a relationship between the number selected and the time of the year in which the man was born. It appears that men who were born between January and June were more likely than expected to have high numbers, whereas men born between July and December were more likely than expected to have low numbers.

5. Because the p-value for this test, 0.0959, is greater than the level of significance $\alpha = 0.05$ (or the computed F test statistic = 2.1913 is less than the critical value of $F = 2.725$), you cannot reject the null hypothesis. You conclude that there is insufficient evidence of a difference in the mean cereal weights among the four plants.

6. Because the p-value for this test, 0.922, is greater than the level of significance $\alpha = 0.05$ (or the computed F test statistic 0.0817 is less than the critical value of $F = 3.6823$), you cannot reject the null hypothesis. You conclude that there is insufficient evidence of a difference in the mean yield between the three methods used in the cleansing step. In other words, there is insufficient evidence of an effect due to the cleansing step.

7. Because the p-value = 0.0000 < 0.05 (or $F = 53.03 > 2.92$), reject H_0. There is evidence of a difference in the mean distances traveled by the golf balls with different designs.

References

1. Berenson, M. L., D. M. Levine, and T. C. Krehbiel. *Basic Business Statistics: Concepts and Applications*, Eleventh Edition. Upper Saddle River, NJ: Prentice Hall, 2009.

2. Conover, W. J. *Practical Nonparametric Statistics*, Third Edition. New York: Wiley, 2000.

3. Daniel, W. *Applied Nonparametric Statistics*, Second Edition. Boston: Houghton Mifflin, 1990.

4. Levine, D. M. *Statistics for Six Sigma Green Belts Using Minitab and JMP* Upper Saddle River, NJ: Prentice Hall, 2006.

5. Levine, D. M., D. Stephan, T. C. Krehbiel, and M. L. Berenson. *Statistics for Managers Using Microsoft Excel*, Fifth Edition. Upper Saddle River, NJ: Prentice Hall, 2008.

6. Levine, D. M., P. P. Ramsey, and R. K. Smidt. *Applied Statistics for Engineers and Scientists Using Microsoft Excel and Minitab*. Upper Saddle River, NJ: Prentice Hall, 2001.

7. Montgomery, D. C. *Design and Analysis of Experiments*, Sixth Edition. New York: John Wiley, 2005.

8. Kutner, M. H., C. Nachtsheim, J. Neter, and W. Li. *Applied Linear Statistical Models*, Fifth Edition., New York: McGraw-Hill-Irwin, 2005.

9. Microsoft Excel 2007. Redmond, WA: Microsoft Corporation, 2006.

Simple Linear Regression

10.1 Basics of Regression Analysis

10.2 Determining the Simple Linear Regression Equation

10.3 Measures of Variation

10.4 Regression Assumptions

10.5 Residual Analysis

10.6 Inferences About the Slope

10.7 Common Mistakes Using Regression Analysis

Important Equations

One-Minute Summary

Test Yourself

Chapter 7, "Fundamentals of Hypothesis Testing," compared hypothesis testing to the scientific method that states tentative hypotheses about natural phenomena and then studies those hypotheses through investigation and testing. The goal of scientific investigation and testing is to develop statements, laws, and theorems that explain or predict natural phenomena such as the equation $E = mc^2$, which relates to the amount of energy, E, that a quantity of matter, m, contains.

When doing statistical analysis, you also might want to be able to state in an equation how the values of one variable are related to another variable. For example, managers of a growing chain of retail stores might wonder if larger-sized stores generate greater sales, farmers might wonder how the weight of a pumpkin is related to its circumference, and baseball fans might wonder how the number of games a team wins in a season is related to the number of runs the team scores.

Statistical methods known as **regression analysis** enable you to develop such mathematical relationships among variables so that you can predict the value of one variable based on another variable. In this chapter, you learn the basic concepts and principles of simple linear regression analysis as well as the statistical assumptions necessary for performing regression analysis.

10.1   Basics of Regression Analysis

In regression analysis, you develop a model that can be used to predict the values of a **dependent** or **response** variable based on the values of one or more **independent** or **explanatory** variables. In this chapter, you learn about simple linear regression, in which you use a single explanatory numerical variable (such as size of store) to predict a numerical dependent variable (such as store sales). In Chapter 11, "Multiple Regression," you will learn the fundamentals of multiple regression, in which several independent variables are used to predict a numerical dependent variable. Other more complicated models including logistic regression, in which the dependent variable is categorical (see References 2, 3, and 6), are not discussed in this book.

Simple Linear Regression

CONCEPT   The statistical method that uses a straight-line relationship to predict a numerical dependent variable Y from a single numerical independent variable X.

INTERPRETATION   Simple linear regression attempts to discover whether the values of the dependent Y (such as store sales) and the independent X variable (such as the size of the store), when graphed on a scatter plot (see Section 2.2), would suggest a straight-line relationship of the values. The following figure shows the different types of patterns that you could discover when plotting the values of the X and Y variables.

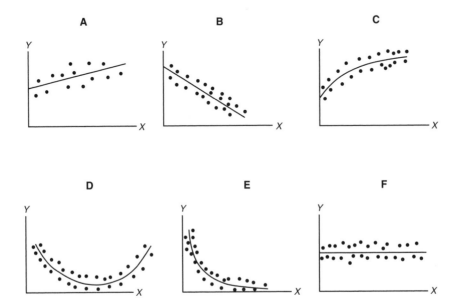

The patterns shown in the preceding figure can be described as follows:

- Panel A, positive straight-line or linear relationship between X and Y.

- Panel B, negative straight-line or linear relationship between X and Y.

- Panel C, a positive curvilinear relationship between X and Y. The values of Y are increasing as X increases, but this increase tapers off beyond certain values of X.

- Panel D, a U-shaped relationship between X and Y. As X increases, at first Y decreases. However, as X continues to increase, Y not only stops decreasing but actually increases above its minimum value.

- Panel E, an exponential relationship between X and Y. In this case, Y decreases very rapidly as X first increases, but then decreases much less rapidly as X increases further.

- Panel F, values that have very little or no relationship between X and Y. High and low values of Y appear at each value of X.

Scatter plots only informally help you identify the relationship between the dependent variable Y and the independent variable X in a simple regression. To specify the numeric relationship between the variables, you need to develop an equation that best represents the relationship.

10.2 Determining the Simple Linear Regression Equation

After you determine that a straight-line relationship exists between a dependent variable Y and the independent variable X, you need to determine which straight line to use to represent the relationship. Two values define any straight line: the Y intercept and the slope.

Y Intercept

CONCEPT   The value of Y when $X = 0$, represented by the symbol b_0.

Slope

CONCEPT   The change in Y per unit change in X represented by the symbol b_1. Positive slope means Y increases as X increases. Negative slope means Y decreases as X increases.

INTERPRETATION The Y intercept and the slope are known as the **regression coefficients**. The symbol b_0 is used for the Y intercept, and the symbol b_1 is used for the slope. Multiplying a specific X value by the slope and then adding the Y intercept generates the corresponding Y value. The equation $Y = b_0 + b_1X$ is used to express this relationship for the entire line. (Some sources use the symbol a for the Y intercept and b for the slope to form the equation $Y = a + bX$.)

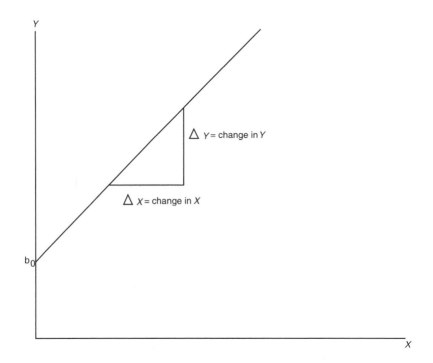

Least-Squares Method

CONCEPT The simple linear regression method that minimizes the sum of the squared differences between the actual values of the dependent variable Y and the predicted values of Y.

INTERPRETATION For plotted sets of X and Y values, there are many possible straight lines, each with its own values of b_0 and b_1, that might seem to fit the data. The least-squares method finds the values for the Y intercept and the slope that makes the sum of the squared differences between the actual values of the dependent variable Y and the predicted values of Y as small as possible.

Calculating the Y intercept and the slope using the least-squares method is tedious and can be subject to rounding errors if you use a simple four-function calculator. You can get more accurate results faster if you use regression software routines to perform the calculations.

WORKED-OUT PROBLEM 1 You want to assist a moving company owner to develop a more accurate method of predicting the labor hours needed for a moving job by using the volume of goods (in cubic feet) that is being moved. The manager has collected the following data for 36 moves and has eliminated the travel-time portion of the time needed for the move.

Moving

Hours	Feet	Hours	Feet
24.00	545	25.00	557
13.50	400	45.00	1,028
26.25	562	29.00	793
25.00	540	21.00	523
9.00	220	22.00	564
20.00	344	16.50	312
22.00	569	37.00	757
11.25	340	32.00	600
50.00	900	34.00	796
12.00	285	25.00	577
38.75	865	31.00	500
40.00	831	24.00	695
19.50	344	40.00	1,054
18.00	360	27.00	486
28.00	750	18.00	442
27.00	650	62.50	1,249
21.00	415	53.75	995
15.00	275	79.50	1,397

The scatter plot for these data (shown on the next page) indicates an increasing relationship between cubic feet moved (X) and labor hours (Y). As the cubic footage moved increases, labor hours increase approximately as a straight line.

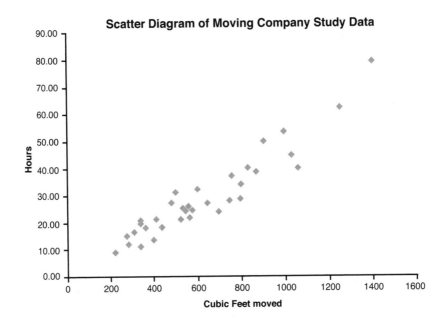

Spreadsheet and calculator regression results for this study are as follows:

	A	B	C	D	E	F	G	H	I
1	Regression Analysis for Moving Company Study								
2									
3	Regression Statistics								
4	Multiple R	0.9430							
5	R Square	0.8892							
6	Adjusted R Square	0.8860							
7	Standard Error	5.0314							
8	Observations	36							
9									
10	ANOVA								
11		df	SS	MS	F	Significance F			
12	Regression	1	6910.7189	6910.7189	272.9864	0.0000			
13	Residual	34	860.7186	25.3153					
14	Total	35	7771.4375						
15									
16		Coefficients	Standard Error	t Stat	P-value	Lower 95%	Upper 95%	Lower 95.0%	Upper 95.0%
17	Intercept	-2.3697	2.0733	-1.1430	0.2610	-6.5830	1.8437	-6.5830	1.84371
18	Cubic Feet Moved	0.0501	0.0030	16.5223	0.0000	0.0439	0.0562	0.0439	0.05624

```
LinRegTTest
y=a+bx
β≠0 and ρ≠0
t=16.52229869
p=8.149363ᴇ-18
df=34
↓a=-2.369660125
```

```
LinRegTTest
y=a+bx
β≠0 and ρ≠0
↑b=.0500802737
s=5.031426629
r²=.8892458914
r=.9429983517
```

The results show that $b_1 = 0.05$ and $b_0 = -2.37$. Thus, the equation for the best straight line for these data is this:

Predicted value of labor hours = $-2.37 + 0.05 \times$ Cubic feet moved

The slope b_1 was computed as +0.05. This means that for each increase of 1 unit in X, the value of Y is estimated to increase by 0.05 units. In other words, for each increase of 1 cubic foot to be moved, the fitted model predicts that the labor hours are estimated to increase by 0.05 hours.

The Y intercept b_0 was computed to be –2.37. The Y intercept represents the value of Y when X equals 0. Because the cubic feet moved cannot be 0, the Y intercept has no practical interpretation. The sample linear regression line for these data, plotted with the actual values, is:

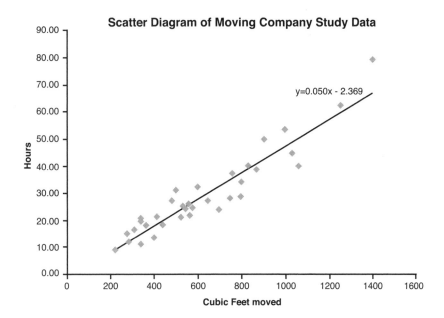

Regression Model Prediction

Once developed, you can use a regression model for predicting values of a dependent variable Y from the independent variable X. However, you are restricted to the relevant range of the independent variable in making predictions. This range is all the values from the smallest to the largest X used to develop the regression model. You should not extrapolate beyond the range of X values. For example, when you use the model developed in WORKED-OUT PROBLEM 1, predictions of labor hours should be made only for moves that are between 220 and 1,397 cubic feet.

WORKED-OUT PROBLEM 2   Using the regression model developed in WORKED-OUT PROBLEM 1, you want to predict the labor hours for a moving job that consists of 800 cubic feet. You predict that the labor hours for a move would be 37.69 (–2.3697 + 0.0501 × 800).

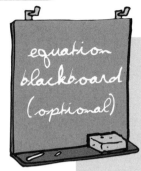

You use the symbols for the Y intercept, b_0, and the slope, b_1, the sample size, n, and these symbols:

- the subscripted YHat, $\hat{Y}_i$, for predicted Y values
- the subscripted italic capital X for the independent X values
- the subscripted italic capital Y for the dependent Y values
- $\bar{X}$ for the mean of the X values
- $\bar{Y}$ for the mean of the Y values

To write the equation for a simple linear regression model:

$$\hat{Y}_i = b_0 + b_1 X_i$$

You use this equation and these summations:

- $\displaystyle\sum_{i=1}^{n} X_i$, the sum of the X values

- $\displaystyle\sum_{i=1}^{n} Y_i$, the sum of the Y values

- $\displaystyle\sum_{i=1}^{n} X_i^2$, the sum of the squared X values

- $\displaystyle\sum_{i=1}^{n} X_i Y_i$, the sum of the cross product of X and Y.

To define the equation of the slope, b_1, as

$$b_1 = \frac{SSXY}{SSX}$$

in which

$$SSXY = \sum_{i=1}^{n}(X_i - \bar{X})(Y_i - \bar{Y}) = \sum_{i=1}^{n} X_i Y_i - \frac{\left(\displaystyle\sum_{i=1}^{n} X_i\right)\left(\displaystyle\sum_{i=1}^{n} Y_i\right)}{n}$$

and

$$SSX = \sum_{i=1}^{n}(X_i - \bar{X})^2 = \sum_{i=1}^{n} X_i^2 - \frac{\left(\displaystyle\sum_{i=1}^{n} X_i\right)^2}{n}$$

These equations, in turn, allow you to define the Y intercept as

$$b_0 = \bar{Y} - b_1 \bar{X}$$

For the moving company example, these sums and the sum of the squared Y values ($\sum_{i=1}^{n} Y_i^2$,) used for calculating the sum of squares total (SST) on page 219 are as follows:

Move	Hours (Y)	Cubic Feet Moved (X)	X^2	Y^2	XY
1	24	545	297,025	576	13,080
2	13.5	400	160,000	182.25	5,400
3	26.25	562	315,844	689.0625	14,752.50
4	25	540	291,600	625	13,500
5	9	220	48,400	81	1,980
6	20	344	118,336	400	6,880
7	22	569	323,761	484	12,518
8	11.25	340	115,600	126.5625	3,825
9	50	900	810,000	2,500	45,000
10	12	285	81,225	144	3,420
11	38.75	865	748,225	1,501.56	33,518.75
12	40	831	690,561	1,600	33.240
13	19.5	344	118,336	380.25	6,708
14	18	360	129,600	324	6,480
15	28	750	562,500	784	21,000
16	27	650	422,500	729	17,550
17	21	415	172,225	441	8,715
18	15	275	75,625	225	4,125
19	25	557	310,249	625	13,925
20	45	1028	1,056,784	2,025	46,260
21	29	793	628,849	841	22,997
22	21	523	273,529	441	10,983
23	22	564	318,096	484	12,408
24	16.5	312	97,344	272.25	5,148
25	37	757	573,049	1,369	28,009

(continues)

Move	Hours (Y)	Cubic Feet Moved (X)	X^2	Y^2	XY
26	32	600	360,000	1,024	19,200
27	34	796	633,616	1,156	27,064
28	25	577	332,929	625	14,425
29	31	500	250,000	961	15,500
30	24	695	483,025	576	16,680
31	40	1054	1,110,916	1,600	42,160
32	27	486	236,196	729	13,122
33	18	442	195,364	324	7,956
34	62.5	1249	1,560,001	3,906.25	78,062.50
35	53.75	995	990,025	2,889.06	53,481.25
36	79.5	1397	1,951,609	6,320.25	111,061.50
Sums: 1,042.5		22,520	16,842,944	37,960.50	790,134.50

Using these sums, you can compute the values of the slope b_1:

$$SSXY = \sum_{i=1}^{n}(X_i - \overline{X})(Y_i - \overline{Y}) = \sum_{i=1}^{n} X_i Y_i - \frac{\left(\sum_{i=1}^{n} X_i\right)\left(\sum_{i=1}^{n} Y_i\right)}{n}$$

$$SSXY = 790,134.5 - \frac{(22,520)(1,042.5)}{36}$$

$$= 790,134.5 - 652,141.66$$

$$= 137,992.84$$

$$SSX = \sum_{i=1}^{n}(X_i - \overline{X})^2 = \sum_{i=1}^{n} X_i^2 - \frac{\left(\sum_{i=1}^{n} X_i\right)^2}{n}$$

$$= 16,842,944 - \frac{(22,520)^2}{36}$$

$$= 16,842,944 - 14,087,511.11$$

$$= 2,755,432.889$$

Because $b_1 = \dfrac{SSXY}{SSX}$

$$b_1 = \dfrac{137,992.84}{2,755,432.889}$$

$$= 0.05008$$

With the value for the slope b_1, you can calculate the Y intercept as follows:

First calculate the mean Y $(\bar{Y})$ and the mean X $(\bar{X})$ values:

$$\bar{Y} = \dfrac{\sum_{i=1}^{n} Y_i}{n} = \dfrac{1,042.5}{36} = 28.9583$$

$$\bar{X} = \dfrac{\sum_{i=1}^{n} X_i}{n} = \dfrac{22,520}{36} = 625.5555$$

Then use these results in the equation

$$b_0 = \bar{Y} - b_1 \bar{X}$$

$$b_0 = 28.9583 - (0.05008)(625.5555)$$

$$= -2.3695$$

10.3 Measures of Variation

After a regression model has been fit to a set of data, three measures of variation determine how much of the variation in the dependent variable Y can be explained by variation in the independent variable X.

Regression Sum of Squares (*SSR*)

CONCEPT   The variation that is due to the relationship between X and Y.

INTERPRETATION   The regression sum of squares (*SSR*) is equal to the sum of the squared differences between the Y values that are predicted from the regression equation and the mean value of Y:

$$SSR = \text{Sum (Predicted } Y \text{ value} - \text{Mean } Y \text{ value)}^2$$

Error Sum of Squares (*SSE*)

CONCEPT   The variation that is due to factors other than the relationship between X and Y.

INTERPRETATION   The error sum of squares (*SSE*) is equal to the sum of the squared differences between each observed Y value and the predicted value of Y:

$$SSE = \text{Sum (Observed } Y \text{ value} - \text{Predicted } Y \text{ value})^2$$

Total Sum of Squares (*SST*)

CONCEPT   The measure of variation of the Y_i values around their mean.

INTERPRETATION   The total sum of squares (*SST*) is equal to the sum of the squared differences between each observed Y value and the mean value of Y:

$$SST = \text{Sum (Observed } Y \text{ value} - \text{Mean } Y \text{ value})^2$$

The total sum of squares is also equal to the sum of the regression sum of squares and the error sum of squares. For the WORKED-OUT PROBLEM of the previous section, *SSR* is 6,910.7189, *SSE* (called residual) is 860.7186, and *SST* is 7,771.4375. (Observe that 7,771.4375 is the sum of 6,910.7189 and 860.7186.)

equation
blackboard
(optional)

interested
in
math?

You use symbols introduced earlier in this chapter to write the equations for the measures of variation in a regression analysis.

The equation for total sum of squares *SST* can be expressed in two ways:

$$SST = \sum_{i=1}^{n}(Y_i - \bar{Y})^2 \text{ which is equivalent to} \sum_{i=1}^{n} Y_i^2 - \frac{\left(\sum_{i=1}^{n} Y_i\right)^2}{n}$$

or as

$$SST = SSR + SSE$$

The equation for the regression sum of squares (SSR) is

SSR = explained variation or regression sum of squares

$$SSR = \sum_{i=1}^{n}(\hat{Y}_i - \bar{Y})^2$$

which is equivalent to

$$= b_0 \sum_{i=1}^{n} Y_i + b_1 \sum_{i=1}^{n} X_i Y_i - \frac{\left(\sum_{i=1}^{n} Y_i\right)^2}{n}$$

The equation for the error sum of squares (SSE) is

SSE = unexplained variation or error sum of squares

$$= \sum_{i=1}^{n}(Y_i - \hat{Y}_i)^2 \text{ which is equivalent to}$$

$$= \sum_{i=1}^{n} Y_i^2 - b_0 \sum_{i=1}^{n} Y_i - b_1 \sum_{i=1}^{n} X_i Y_i$$

For the moving company example on page 211,

$$SST = \text{total sum of squares} = \sum_{i=1}^{n}(Y_i - \bar{Y})^2 = \sum_{i=1}^{n} Y_i^2 - \frac{\left(\sum_{i=1}^{n} Y_i\right)^2}{n}$$

$$= 37,960.5 - \frac{(1,042.5)^2}{36}$$

$$= 37,960.5 - 30,189.0625$$

$$= 7,771.4375$$

SSR = regression sum of squares

$$= \sum_{i=1}^{n}(\hat{Y}_i - \bar{Y})^2$$

$$= b_0 \sum_{i=1}^{n} Y_i + b_1 \sum_{i=1}^{n} X_i Y_i - \frac{\left(\sum_{i=1}^{n} Y_i\right)^2}{n}$$

$$= (-2.3695)(1,042.5) + (0.05008)(790,134.5) - \frac{(1,042.5)^2}{36}$$

$$= 6,910.671$$

(continues)

SSE = error sum of squares

$$= \sum_{i=1}^{n}(Y_i - \hat{Y}_i)^2$$

$$= \sum_{i=1}^{n} Y_i^2 - b_0 \sum_{i=1}^{n} Y_i - b_1 \sum_{i=1}^{n} X_i Y_i$$

$$= 37,960.5 - (-2.3695)(1,042.5) - (0.05008)(790,134.5)$$

$$= 860.768$$

Calculated as $SSR + SSE$, the total sum of squares SST is 7,771.439, slightly different from the results from the first equation because of rounding errors.

The Coefficient of Determination

CONCEPT   The ratio of the regression sum of squares to the total sum of squares, represented by the symbol r^2.

INTERPRETATION   By themselves, SSR, SSE, and SST provide little that can be directly interpreted. The ratio of the regression sum of squares (SSR) to the total sum of squares (SST) measures the proportion of variation in Y that is explained by the independent variable X in the regression model. The ratio can be expressed as follows:

$$r^2 = \frac{\text{regression sum of squares}}{\text{total sum of squares}} = \frac{SSR}{SST}$$

For the moving company example, SSR = 6,910.7189 and SST = 7,771.4375 (see regression results on page 212). Therefore

$$r^2 = \frac{6,910.719}{7,771.4375} = 0.8892$$

This result for r^2 means that 89% of the variation in labor hours can be explained by the variability in the cubic footage to be moved. This shows a strong positive linear relationship between the two variables, because the use of a regression model has reduced the variability in predicting labor hours by 89%. Only 11% of the sample variability in labor hours can be explained by factors other than what is accounted for by the linear regression model that uses only cubic footage.

The Coefficient of Correlation

CONCEPT   The measure of the strength of the linear relationship between two variables, represented by the symbol r.

INTERPRETATION   The values of this coefficient vary from -1, which indicates perfect negative correlation, to $+1$, which indicates perfect positive correlation. The sign of the correlation coefficient r is the same as the sign of the slope in simple linear regression. If the slope is positive, r is positive. If the slope is negative, r is negative. The coefficient of correlation (r) is the square root of the coefficient of determination r^2.

For the moving company example, the coefficient of correlation, r, is $+0.943$, the positive (since the slope is positive) square root of 0.8892 (r^2). (Microsoft Excel labels the coefficient of correlation as "multiple r.") Because the coefficient is very close to $+1.0$, you can say that the relationship between cubic footage moved and labor hours is very strong. You can plausibly conclude that the increased volume that had to be moved is associated with increased labor hours.

important point

In general, you must remember that just because two variables are strongly correlated, you cannot conclude that a cause-and-effect relationship exists between the variables.

Standard Error of the Estimate

CONCEPT   The standard deviation around the fitted line of regression that measures the variability of the actual Y values from the predicted Y, represented by the symbol S_{YX}.

INTERPRETATION   Although the least-squares method results in the line that fits the data with the minimum amount of variation, unless the coefficient of determination $r^2 = 1.0$, the regression equation is not a perfect predictor.

The variability around the line of regression was shown in the figure on page 212, which presented the scatter plot and the line of regression for the moving company data. You can see from that figure that some values are above the line of regression and other values are below the line of regression. For the moving company example, the standard error of the estimate (labeled as Standard Error in the figure on page 212) is equal to 5.03 hours.

Just as the standard deviation measures variability around the mean, the standard error of the estimate measures variability around the fitted line of regression. As you will see in Section 10.6, the standard error of the estimate can be used to determine whether a statistically significant relationship exists between the two variables.

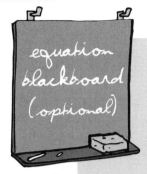

You use symbols introduced earlier in this chapter to write the equation for the standard error of the estimate:

$$S_{YX} = \sqrt{\frac{SSE}{n-2}} = \sqrt{\frac{\sum_{i=1}^{n}(Y_i - \hat{Y}_i)^2}{n-2}}$$

For the moving company problem, with SSE equal to 860.7186,

$$S_{YX} = \sqrt{\frac{860.7186}{36-2}}$$

$$S_{YX} = 5.0314$$

10.4   Regression Assumptions

The assumptions necessary for performing a regression analysis are as follows:

- Normality of the variation around the line of regression
- Equality of variation in the Y values for all values of X
- Independence of the variation around the line of regression

The first assumption, normality, requires that the variation around the line of regression is normally distributed at each value of X. Like the t test and the ANOVA F test, regression analysis is fairly insensitive to departures from the normality assumption. As long as the distribution of the variation around the line of regression at each level of X is not extremely different from a normal distribution, inferences about the line of regression and the regression coefficients will not be seriously affected.

The second assumption, equality of variation, requires that the variation around the line of regression be constant for all values of X. This means that the variation is the same when X is a low value as when X is a high value. The equality of variation assumption is important for using the least-squares method of determining the regression coefficients. If there are serious departures from this assumption, other methods (see Reference 4) can be used.

The third assumption, independence of the variation around the line of regression, requires that the variation around the regression line be independent for each value of X. This assumption is particularly important when data are collected over a period of time. In such situations, the variation around the line for a specific time period is often correlated with the variation of the previous time period.

10.5 Residual Analysis

The graphical method, **residual analysis**, enables you to evaluate whether the regression model that has been fitted to the data is an appropriate model and determine whether there are violations of the assumptions of the regression model.

Residual

CONCEPT   The difference between the observed and predicted values of the dependent variable Y for a given value of X.

INTERPRETATION   To evaluate the aptness of the fitted model, you plot the residuals on the vertical axis against the corresponding X values of the independent variable on the horizontal axis. If the fitted model is appropriate for the data, there will be no apparent pattern in this plot. However, if the fitted model is not appropriate, there will be a clear relationship between the X values and the residuals.

A residual plot for the moving company data fitted line of regression appears in the following figure. In this figure, the cubic feet are plotted on the horizontal X axis and the residuals are plotted on the vertical Y axis. You see that although there is widespread scatter in the residual plot, no apparent pattern or relationship exists between the residuals and X. The residuals appear to be evenly spread above and below 0 for the differing values of X. This result enables you to conclude that the fitted straight-line model is appropriate for the moving company data.

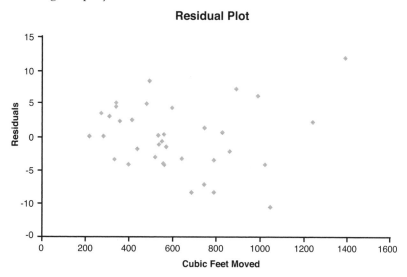

Residual Plot

Evaluating the Assumptions

Different techniques, all involving the residuals, enable you to evaluate the regression assumptions.

For equality of variation, you use the same plot as you did to evaluate the aptness of the fitted model. For the moving company residual plot shown in the preceding figure, there do not appear to be major differences in the variability of the residuals for different X values. You can conclude that for this fitted model, there is no apparent violation in the assumption of equal variation at each level of X.

For the normality of the variation around the line of regression, you plot the residuals in a histogram (see Section 2.2), box-and-whisker plot (see Section 3.4), or a normal probability plot (see Section 5.4). From the histogram shown in the following figure for the moving company data, you can see that the data appear to be approximately normally distributed, with most of the residuals concentrated in the center of the distribution.

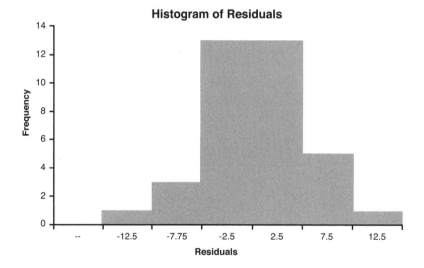

For the independence of the variation around the line of regression, you plot the residuals in the order or sequence in which the observed data was obtained, looking for a relationship between consecutive residuals. If you can see such a relationship, the assumption of independence is violated. Because these data were not collected over time, you do not need to evaluate this assumption.

10.6 Inferences About the Slope

You can make inferences about the linear relationship between the variables in a population based on your sample results after using residual analysis to determine whether the assumptions of the least-squares regression model have not been seriously violated and that the straight-line model is appropriate.

t Test for the Slope

You can determine the existence of a significant relationship between the X and Y variables by testing whether β_1 (the population slope) is equal to 0. If this hypothesis is rejected, you conclude that evidence of a linear relationship exists. The null and alternative hypotheses are as follows:

$$H_0\text{: } \beta_1 = 0 \quad \text{(No linear relationship exists.)}$$

$$H_1\text{: } \beta_1 \neq 0 \quad \text{(A linear relationship exists.)}$$

The test statistic follows the t distribution with the degrees of freedom equal to the sample size minus 2. The test statistic is equal to the sample slope divided by the standard error of the slope:

$$t = \frac{\text{sample slope}}{\text{standard error of the slope}}$$

For the moving company example (see the results on page 212), the critical value of t at the level of significance of $\alpha = 0.05$ is 2.0322, the value of t is 16.52, and the p-value is 0.0000. (Microsoft Excel labels the t statistic "t Stat" on page 212.) Using the p-value approach, you reject H_0 because the p-value of 0.0000 is less than $\alpha = 0.05$. Using the critical value approach, you reject H_0 because $t = 16.52 > 2.0322$. You can conclude that a significant linear relationship exists between labor hours and the cubic footage moved.

equation blackboard (optional)

interested in math?

You assemble symbols introduced earlier and the symbol for the standard error of the slope, S_{b_1}, to form the equation for the t statistic used in testing a hypothesis for a population slope β_1.

You begin by forming the equations for the standard error of the slope, S_{b_1} as

$$S_{b_1} = \frac{S_{YX}}{\sqrt{SSX}}$$

(continues)

interested in math?

Then, you use the standard error of the slope, S_{b_1} to define the test statistic:

$$t = \frac{b_1 - \beta_1}{S_{b_1}}$$

The test statistic t follows a t distribution with $n - 2$ degrees of freedom.

For the moving company example, to test whether a significant relationship exists between the cubic footage and the labor hours at the level of significance $\alpha = 0.05$, refer to the calculation of SSX on page 216 and the standard error of the estimate on page 222.

$$S_{b_1} = \frac{S_{YX}}{\sqrt{SSX}}$$

$$= \frac{5.0314}{\sqrt{2,755,432.889}}$$

$$= 0.00303$$

Therefore, to test the existence of a linear relationship at the 0.05 level of significance, with

$$b_1 = +0.05008 \quad n = 36 \quad S_{b_1} = 0.00303$$

$$t = \frac{b_1 - \beta_1}{S_{b_1}}$$

$$= \frac{0.05008 - 0}{0.00303} = 16.52$$

Confidence Interval Estimate of the Slope (β_1)

You can also test the existence of a linear relationship between the variables by constructing a confidence interval estimate of β_1 and determining whether the hypothesized value ($\beta_1 = 0$) is included in the interval.

You construct the confidence interval estimate of the slope β_1 by multiplying the t statistic by the standard error of the slope and then adding and subtracting this product to the sample slope.

For the moving company example, the regression results on page 212 include the calculated lower and upper limits of the confidence interval estimate for the slope of cubic footage and labor hours. With 95% confidence, the lower limit is 0.0439 and the upper limit is 0.0562.

Because these values are above 0, you conclude that a significant linear relationship exists between labor hours and cubic footage moved. The confidence interval indicates that for each increase of 1 cubic foot moved, the mean labor hours are estimated to increase by at least 0.0439 hours but less than 0.0562 hours. Had the interval included 0, you would have concluded that no relationship exists between the variables.

equation blackboard (optional)

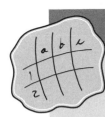

interested in math?

You assemble symbols introduced earlier to form the equation for the confidence interval estimate of the slope β_1:

$$b_1 \pm t_{n-2} S_{b_1}$$

For the moving company example, b_1 has already been calculated on page 217 and the standard error of the slope S_{b_1} has already been calculated on page 226.

$$b_1 = +0.05008 \quad n = 36 \quad S_{b_1} = 0.00303$$

Thus, using 95% confidence, with degrees of freedom = $36 - 2 = 34$,

$$b_1 \pm t_{n-2} S_{b_1}$$

$$= +0.05008 \pm (2.0322)(0.00303)$$

$$= +0.05008 \pm 0.0061$$

$$+0.0439 \le \beta_1 \le +0.0562$$

spreadsheet solution

Simple Linear Regression

Chapter 10 Simple Linear Regression ATP (shown on page 212) contains the regression results for the moving company study data as calculated by the Analysis ToolPak **Regression** procedure. Spreadsheet Tip ATT6 in Appendix D explains how to use the **Regression** procedure.

Alternative method. If you use OpenOffice.org Calc 3, you must use the **LINEST** function, discussed in Section E.4 of Appendix E to calculate regression results. (You can also use the **LINEST** function in Microsoft Excel versions 2003 or later.)

calculator keys

Simple Linear Regression

Enter the data values for your dependent variable Y and your independent variable X into separate list variables (see Chapter 1). Then, press [STAT][◀] (to display the TESTS menu) and select F:LinRegTTest(and press [ENTER] to display the LinRegTTest function. (Note: In TI-83 series calculators, select E:LinRegTTest(to display this function.)

In the function screen:

1. Enter the list variable of your independent variable as **Xlist** and the list variable of your dependent variable as **Ylist**.

2. Enter 1 as the **Freq**.

3. For **RegEQ** entry, press [VARS][◀] (to display the Y-VARS menu) and then select **1:Function** and press [ENTER]. In the FUNCTION screen that appears, select **1:Y1** and press [ENTER]. (This assigns the special built-in function variable **Y1** to **RegEQ**.)

4. Select **Calculate** and press [ENTER].

Regression results appear over two screens. If your two list variables do not have the same number of values, an error message appears and you need to quit the procedure.

See the **Calculator Keys** section in Chapter 11 for an alternative calculator regression analysis method.

10.7 Common Mistakes Using Regression Analysis

Some of the common mistakes that people make when using regression analysis are as follows:

important point

- Lacking an awareness of the assumptions of least-squares regression
- Knowing how to evaluate the assumptions of least-squares regression
- Knowing what the alternatives to least-squares regression are if a particular assumption is violated

- Using a regression model without knowledge of the subject matter
- Predicting Y outside the relevant range of X

Most software regression analysis routines do not check for these mistakes. You must always use regression analysis wisely and always check that others who provide you with regression results have avoided these mistakes as well.

The following four sets of data illustrate some of the mistakes that you can make in a regression analysis.

Anscombe

Data Set A		Data Set B		Data Set C		Data Set D	
X_i	Y_i	X_i	Y_i	X_i	Y_i	X_i	Y_i
10	8.04	10	9.14	10	7.46	8	6.58
14	9.96	14	8.10	14	8.84	8	5.76
5	5.68	5	4.74	5	5.73	8	7.71
8	6.95	8	8.14	8	6.77	8	8.84
9	8.81	9	8.77	9	7.11	8	8.47
12	10.84	12	9.13	12	8.15	8	7.04
4	4.26	4	3.10	4	5.39	8	5.25
7	4.82	7	7.26	7	6.42	19	12.50
11	8.33	11	9.26	11	7.81	8	5.56
13	7.58	13	8.74	13	12.74	8	7.91
6	7.24	6	6.13	6	6.08	8	6.89

Source: Extracted from F. J. Anscombe, "Graphs in Statistical Analysis," *American Statistician*, Vol. 27 (1973), pp. 17–21.

Anscombe (Reference 1) showed that for the four data sets, the regression results are identical:

$$\text{predicted value of } Y = 3.0 + 0.5X_i$$

$$\text{standard error of the estimate} = 1.237$$

$$r^2 = 0.667$$

$$SSR = \text{regression sum of squares} = 27.51$$

$$SSE = \text{error sum of squares} = 13.76$$

$$SST = \text{total sum of squares} = 41.27$$

However, the four data sets are actually quite different as scatter plots and residual plots for the four sets reveal.

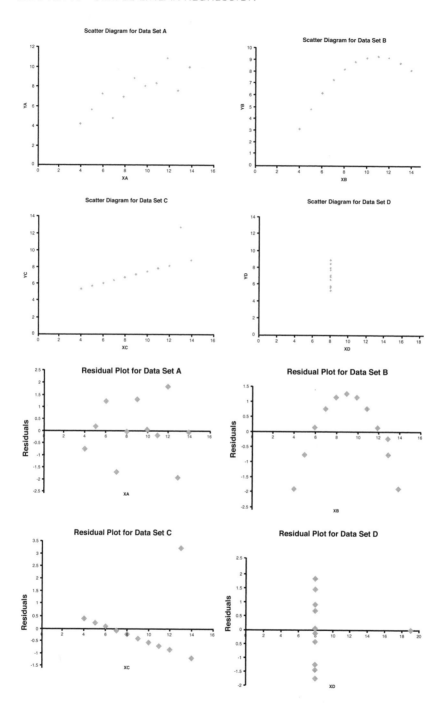

From the scatter plots and the residual plots, you see how different the data sets are. The only data set that seems to follow an approximate straight line is data set *A*. The residual plot for data set *A* does not show any obvious patterns or outlying residuals. This is certainly not true for data sets *B*, *C*, and *D*.

The scatter plot for data set *B* shows that a curvilinear regression model should be considered. The residual plot reinforces this conclusion for *B*. The scatter plot and the residual plot for data set *C* clearly depict an extreme value. Similarly, the scatter plot for data set *D* represents the unusual situation in which the fitted model is heavily dependent on the outcome of a single observation ($X = 19$ and $Y = 12.50$). Any regression model fit for these data should be evaluated cautiously, because its regression coefficients are heavily dependent on a single observation.

To avoid the common mistakes of regression analysis, you can use the following process:

- Always start with a scatter plot to observe the possible relationship between *X* and *Y*.

- Check the assumptions of regression after the regression model has been fit, before using the results of the model.

- Plot the residuals versus the independent variable. This enables you to determine whether the model fit to the data is an appropriate one and enables you to check visually for violations of the equal variation assumption.

- Use a histogram, box-and-whisker plot, or normal probability plot of the residuals to graphically evaluate whether the normality assumption has been seriously violated.

- If the evaluation of the residuals indicates violations in the assumptions, use alternative methods to least-squares regression or alternative least-squares models (see Reference 4), depending on what the evaluation has indicated.

- If the evaluation of the residuals does not indicate violations in the assumptions, then you can undertake the inferential aspects of the regression analysis. You can conduct a test for the significance of the slope and you can construct a confidence interval estimate of the slope.

Important Equations

Regression equation

(10.1) $\hat{Y}_i = b_0 + b_1 X_i$

Slope

(10.2) $b_1 = \dfrac{SSXY}{SSX}$

(10.3)
$$SSXY = \sum_{i=1}^{n}(X_i - \overline{X})(Y_i - \overline{Y}) = \sum_{i=1}^{n} X_i Y_i - \dfrac{\left(\sum_{i=1}^{n} X_i\right)\left(\sum_{i=1}^{n} Y_i\right)}{n}$$

and

$$SSX = \sum_{i=1}^{n}(X_i - \overline{X})^2 = \sum_{i=1}^{n} X_i^2 - \dfrac{\left(\sum_{i=1}^{n} X_i\right)^2}{n}$$

Y intercept

(10.4) $b_0 = \overline{Y} - b_1 \overline{X}$

Total Sum of Squares

(10.5) $SST = \sum_{i=1}^{n}(Y_i - \overline{Y})^2$ which is equivalent to $\sum_{i=1}^{n} Y_i^2 - \dfrac{\left(\sum_{i=1}^{n} Y_i\right)^2}{n}$

(10.6) $SST = SSR + SSE$

Regression Sum of Squares

(10.7)
SSR = explained variation or regression sum of squares

$= \sum_{i=1}^{n}(\hat{Y}_i - \overline{Y})^2$

which is equivalent to

$= b_0 \sum_{i=1}^{n} Y_i + b_1 \sum_{i=1}^{n} X_i Y_i - \dfrac{\left(\sum_{i=1}^{n} Y_i\right)^2}{n}$

Error Sum of Squares

(10.8) SSE = unexplained variation or error sum of squares

$$= \sum_{i=1}^{n}(Y_i - \hat{Y}_i)^2 \text{ which is equivalent to}$$

$$= \sum_{i=1}^{n} Y_i^2 - b_0 \sum_{i=1}^{n} Y_i - b_1 \sum_{i=1}^{n} X_i Y_i$$

Coefficient of Determination

(10.9) $r^2 = \dfrac{\text{regression sum of squares}}{\text{total sum of squares}} = \dfrac{SSR}{SST}$

Coefficient of Correlation

(10.10) $r = \sqrt{r^2}$ If b_1 is positive, r is positive. If b_1 is negative, r is negative.

Standard Error of the Estimate

(10.11) $S_{YX} = \sqrt{\dfrac{SSE}{n-2}} = \sqrt{\dfrac{\sum_{i=1}^{n}(Y_i - \hat{Y}_i)^2}{n-2}}$

t Test for the Slope

(10.12) $t = \dfrac{b_1 - \beta_1}{S_{b_1}}$

One-Minute Summary

Simple Linear Regression

- Least-squares method
- Measures of variation
- Residual analysis
- t test for the significance of the slope
- Confidence interval estimate of the slope

Test Yourself

Short Answers

1. The Y intercept (b_0) represents the:
   (a) predicted value of Y when $X = 0$

   (b) change in Y per unit change in X

   (c) predicted value of Y

   (d) variation around the regression line

2. The slope (b_1) represents:
   (a) predicted value of Y when $X = 0$

   (b) change in Y per unit change in X

   (c) predicted value of Y

   (d) variation around the regression line

3. The standard error of the estimate is a measure of:
   (a) total variation of the Y variable

   (b) the variation around the regression line

   (c) explained variation

   (d) the variation of the X variable

4. The coefficient of determination (r^2) tells you:
   (a) that the coefficient of correlation (r) is larger than 1

   (b) whether the slope has any significance

   (c) whether the regression sum of squares is greater than the total sum of squares

   (d) the proportion of total variation that is explained

5. In performing a regression analysis involving two numerical variables, you assume:
   (a) the variances of X and Y are equal

   (b) the variation around the line of regression is the same for each X value

   (c) that X and Y are independent

   (d) All of the above

6. Which of the following assumptions concerning the distribution of the variation around the line of regression (the residuals) is correct?
   (a) The distribution is normal.

   (b) All of the variations are positive.

   (c) The variation increases as X increases.

   (d) Each residual is dependent on the previous residual.

7. The residuals represent:
    (a) the difference between the actual Y values and the mean of Y
    (b) the difference between the actual Y values and the predicted Y values
    (c) the square root of the slope
    (d) the predicted value of Y when $X = 0$

8. If the coefficient of determination (r^2) = 1.00, then:
    (a) the Y intercept must equal 0
    (b) the regression sum of squares (SSR) equals the error sum of squares (SSE)
    (c) the error sum of squares (SSE) equals 0
    (d) the regression sum of squares (SSR) equals 0

9. If the coefficient of correlation (r) = –1.00, then:
    (a) All the data points must fall exactly on a straight line with a slope that equals 1.00.
    (b) All the data points must fall exactly on a straight line with a negative slope.
    (c) All the data points must fall exactly on a straight line with a positive slope.
    (d) All the data points must fall exactly on a horizontal straight line with a zero slope.

10. Assuming a straight line (linear) relationship between X and Y, if the coefficient of correlation (r) equals –0.30:
    (a) there is no correlation
    (b) the slope is negative
    (c) variable X is larger than variable Y
    (d) the variance of X is negative

11. The strength of the linear relationship between two numerical variables is measured by the:
    (a) predicted value of Y
    (b) coefficient of determination
    (c) total sum of squares
    (d) Y intercept

12. In a simple linear regression model, the coefficient of correlation and the slope:
    (a) may have opposite signs
    (b) must have the same sign
    (c) must have opposite signs
    (d) are equal

Answer True or False:

13. The regression sum of squares (*SSR*) can never be greater than the total sum of squares (*SST*).

14. The coefficient of determination represents the ratio of *SSR* to *SST*.

15. Regression analysis is used for prediction, while correlation analysis is used to measure the strength of the association between two numerical variables.

16. The value of *r* is always positive.

17. When the coefficient of correlation $r = -1$, a perfect relationship exists between *X* and *Y*.

18. If no apparent pattern exists in the residual plot, the regression model fit is appropriate for the data.

19. If the range of the *X* variable is between 100 and 300, you should not make a prediction for *X* = 400.

20. If the *p*-value for a *t* test for the slope is 0.021, the results are significant at the 0.01 level of significance.

Fill in the blank:

21. The residual represents the difference between the observed value of *Y* and the _____ value of *Y*.

22. The change in *Y* per unit change in *X* is called the _____.

23. The ratio of the regression sum of squares (*SSR*) to the total sum of squares (*SST*) is called the _____.

24. In simple linear regression, if the slope is positive, then the coefficient of correlation must also be _____.

25. One of the assumptions of regression is that the residuals around the line of regression follow the _____ distribution.

Answers to Test Yourself Short Answers

1. a	10. b
2. b	11. b
3. b	12. b
4. d	13. True
5. b	14. True
6. a	15. True
7. b	16. False
8. c	17. True
9. b	18. True

19. True

20. False

21. predicted

22. slope

23. coefficient of determination

24. positive

25. normal

Problems

1. The appraised value (in thousands of dollars) and land area (in acres) was collected for a sample of 30 single-family homes located in Glen Cove, New York (see the following table). Develop a simple linear regression model to predict appraised value based on the land area of the property.

GlenCove

Appraised Value	Land (acres)	Appraised Value	Land (acres)
466.0	0.2297	288.4	0.1714
364.0	0.2192	396.7	0.3849
429.0	0.1630	613.5	0.6545
548.4	0.4608	314.1	0.1722
405.9	0.2549	363.5	0.1435
374.1	0.2290	364.3	0.2755
315.0	0.1808	305.1	0.1148
749.7	0.5015	441.7	0.3636
217.7	0.2229	353.1	0.1474
635.7	0.1300	463.3	0.2281
350.7	0.1763	320.0	0.4626
455.0	0.4200	332.8	0.1889
356.2	0.2520	276.6	0.1228
271.7	0.1148	397.0	0.1492
304.3	0.1693	221.9	0.0852

(a) Assuming a linear relationship, use the least-squares method to compute the regression coefficients b_0 and b_1. State the regression equation for predicting the appraised value based on the land area.

(b) Interpret the meaning of the Y intercept b_0 and the slope b_1 in this problem.

(c) Explain why the regression coefficient, b_0, has no practical meaning in the context of this problem.

(d) Predict the appraised value for a house that has a land area of 0.25 acres.

(e) Compute the coefficient of determination, r^2, and interpret its meaning.

(f) Perform a residual analysis on the results and determine the adequacy of the model.

(g) Determine whether a significant relationship exists between appraised value and the land area of a property at the 0.05 level of significance.

(h) Construct a 95% confidence interval estimate of the population slope between the appraised value and the land area of a property.

2. Measuring the height of a California redwood tree is a very difficult undertaking because these trees grow to heights of more than 300 feet. People familiar with these trees understand that the height of a California redwood tree is related to other characteristics of the tree, including the diameter of the tree at the breast height of a person. The following data represent the height (in feet) and diameter at breast height of a person for a sample of 21 California redwood trees.

Redwood

Height	Diameter at breast height	Height	Diameter at breast height
122.0	20	164.0	40
193.5	36	203.3	52
166.5	18	174.0	30
82.0	10	159.0	22
133.5	21	205.0	42
156.0	29	223.5	45
172.5	51	195.0	54
81.0	11	232.5	39
148.0	26	190.5	36
113.0	12	100.0	8
84.0	13		

(a) Assuming a linear relationship, use the least-squares method to compute the regression coefficients b_0 and b_1. State the regression equation that predicts the height of a tree based on the tree's diameter at breast height of a person.

(b) Interpret the meaning of the slope in this equation.

(c) Predict the height for a tree that has a breast diameter of 25 inches.

(d) Interpret the meaning of the coefficient of determination in this problem.

(e) Perform a residual analysis on the results and determine the adequacy of the model.

(f) Determine whether a significant relationship exists between the height of redwood trees and the breast diameter at the 0.05 level of significance.

(g) Construct a 95% confidence interval estimate of the population slope between the height of the redwood trees and breast diameter.

3. A baseball analyst would like to study various team statistics for the 2008 baseball season to determine which variables might be useful in predicting the number of wins achieved by teams during the season. He has decided to begin by using a team's earned run average (ERA), a measure of pitching performance, to predict the number of wins. The data for the 30 Major League Baseball teams are as follows:

BB2008

Team	Wins	ERA	Team	Wins	ERA
Baltimore	68	5.13	Atlanta	72	4.46
Boston	95	4.01	Chicago Cubs	97	3.87
Chicago White Sox	88	4.09	Cincinnati	74	4.55
Cleveland	81	4.45	Colorado	74	4.77
Detroit	74	4.90	Florida	84	4.43
Kansas City	75	4.48	Houston	86	4.36
Los Angeles Angels	100	3.99	Los Angeles Dodgers	84	3.68
Minnesota	88	4.18	Milwaukee	90	3.85
New York Yankees	89	4.28	New York Mets	89	4.07
Oakland	75	4.01	Philadelphia	92	3.88
Seattle	61	4.73	Pittsburgh	67	5.08
Tampa Bay	97	3.82	St. Louis	86	4.19
Texas	79	5.37	San Diego	63	4.41
Toronto	86	3.49	San Francisco	72	4.38
Arizona	82	3.98	Washington	59	4.66

(Hint: First, determine which are the independent and dependent variables.)

   (a) Assuming a linear relationship, use the least-squares method to compute the regression coefficients b_0 and b_1. State the regression equation for predicting the number of wins based on the ERA.

   (b) Interpret the meaning of the Y intercept, b_0, and the slope, b_1, in this problem.

   (c) Predict the number of wins for a team with an ERA of 4.50.

   (d) Compute the coefficient of determination, r^2, and interpret its meaning.

   (e) Perform a residual analysis on your results and determine the adequacy of the fit of the model.

   (f) At the 0.05 level of significance, does evidence exist of a linear relationship between the number of wins and the ERA?

   (g) Construct a 95% confidence interval estimate of the slope.

   (h) What other independent variables might you include in the model?

Restaurants

4. Zagat's publishes restaurant ratings for various locations in the United States. For this analysis, a sample of 50 restaurants located in an urban area (New York City) and 50 restaurants located in a suburb of New York City are selected and the Zagat rating for food, decor, service, and the cost per person for each restaurant is recorded. Develop a regression model to predict the cost per person, based on a variable that represents the sum of the ratings for food, decor, and service.

Source: Extracted from *Zagat Survey 2007 New York City Restaurants* and *Zagat Survey 2007–2008, Long Island Restaurants.*

   (a) Assuming a linear relationship, use the least-squares method to compute the regression coefficients b_0 and b_1. State the regression equation for predicting the cost per person based on the summated rating.

   (b) Interpret the meaning of the Y intercept, b_0, and the slope, b_1, in this problem.

   (c) Predict the cost per person for a restaurant with a summated rating of 50.

   (d) Compute the coefficient of determination, r^2, and interpret its meaning.

   (e) Perform a residual analysis on your results and determine the adequacy of the fit of the model.

   (f) At the 0.05 level of significance, does evidence exist of a linear relationship between the cost per person and the summated rating?

   (g) Construct a 95% confidence interval estimate of the mean cost per person for all restaurants with a summated rating of 50.

   (h) How useful do you think the summated rating is as a predictor of cost? Explain.

Answers to Test Yourself Problems

1. (a) $b_0 = 257.2253$, $b_1 = 538.4873$; Predicted appraised value = 257.2253 + 538.4873 acre

   (b) Each increase by one acre in land area is estimated to increase appraised value by \$538.4873 thousands.

   (c) The interpretation of b_0 has no practical meaning here because it would represent the estimated appraised value of a house that has no land area.

   (d) Predicted appraised value = 257.2253 + 538.4873 (0.25) = \$391.85 thousands.

   (e) $r^2 = 0.3770$ 37.70% of the variation in appraised value of a house can be explained by variation in land area.

   (f) There is no particular pattern in the residual plot, and the model appears to be adequate.

   (g) $t = 4.1159$; p-value 0.0003 < 0.05 (or $t = 4.1159 > 2.0484$). Reject H_0 at 5% level of significance. There is evidence of a significant linear relationship between appraised value and land area.

   (h) $270.489 < \beta_1 < 806.4856$

2. (a) $b_0 = 78.7963$, $b_1 = 2.6732$; Predicted height = 78.7963 + 2.6732 diameter of the tree at breast height of a person (in inches).

   (b) For each additional inch in the diameter of the tree at breast height of a person, the height of the tree is estimated to increase by 2.6732 feet.

   (c) Predicted height = 78.7963 + 2.6732 (25) = 145.6267 feet

   (d) $r^2 = 0.7288$. 72.88% of the total variation in the height of the tree can be explained by the variation of the diameter of the tree at breast height of a person.

   (e) There is no particular pattern in the residual plot, and the model appears to be adequate.

   (f) $t = 7.1455$; p-value = virtually 0 < 0.05 ($t = 7.1455 > 2.093$). Reject H_0. There is evidence of a significant linear relationship between the height of the tree and the diameter of the tree at breast height of a person.

   (g) $1.8902 < \beta_1 < 3.4562$

3. (a) $b_0 = 152.4045$, $b_1 = -16.5584$; Predicted wins = 152.4045 − 16.5584 ERA

   (b) For each additional earned run allowed, the number of wins is estimated to decrease by 16.5584.

   (c) Predicted wins = 152.4045 − 16.5584 (4.5) = 77.89

   (d) $r^2 = 0.4508$. 45.08% of the total variation in the number of wins can be explained by the variation of the ERA.

(e) There is no particular pattern in the residual plot, and the model appears to be adequate.

(f) $t = -4.7942$ p-value = virtually $0 < 0.05$ ($t = -4.7942 < -2.0484$). Reject H_0. There is evidence of a significant linear relationship between the number of wins and the ERA.

(g) $-23.6333 < \beta_1 < -9.4835$

(h) Among the independent variables you could consider including in the model are runs scored, hits allowed, saves, walks allowed, and errors. For a discussion of multiple regression models that consider several independent variables, see Chapter 11.

4. (a) $b_0 = -30.2524$, $b_1 = 1.2267$ predicted cost $= -30.2524 + 1.2267$ summated rating

(b) For each additional unit increase in summated rating, the cost per person is estimated to increase by $1.23. Because no restaurant will receive a summated rating of 0, it is inappropriate to interpret the Y intercept.

(c) predicted cost $= -30.2524 + 1.2267 (50) = \$31.08$

(d) $r^2 = 0.5452$. So, 54.52% of the variation in the cost per person can be explained by the variation in the summated rating.

(e) There is no obvious pattern in the residuals so the assumptions of regression are met. The model appears to be adequate.

(f) $t = 10.84$, the p-value is virtually $0 < 0.05$ (or $t = 10.84 > 1.9845$); reject H_0. There is evidence of a linear relationship between cost per person and summated rating.

(g) $1.0021 \leq \beta_1 \leq 1.4513$

(h) The linear regression model appears to have provided an adequate fit and shows a significant linear relationship between price per person and summated rating. Because 54.52% of the variation in the cost per person can be explained by the variation in summated rating, summated rating is moderately useful in predicting the cost per person.

References

1. Anscombe, F. J. "Graphs in Statistical Analysis." *American Statistician* 27 (1973): 17–21.

2. Berenson, M. L., D. M. Levine, and T. C. Krehbiel. *Basic Business Statistics: Concepts and Applications*, Eleventh Edition. Upper Saddle River, NJ: Prentice Hall, 2009.

3. Hosmer, D. W., and S. Lemeshow, *Applied Logistic Regression*, 2nd ed. (New York: Wiley, 2001).

4. Kutner, M. H., C. Nachtsheim, J. Neter, and W. Li. *Applied Linear Statistical Models* 5th Ed. (New York: McGraw-Hill-Irwin, 2005).

5. Levine, D. M., D. Stephan, T. C. Krehbiel, and M. L. Berenson. *Statistics for Managers Using Microsoft Excel*, Fifth Edition. Upper Saddle River, NJ: Prentice Hall, 2008.

6. Levine, D. M., P. P. Ramsey, and R. K. Smidt. *Applied Statistics for Engineers and Scientists Using Microsoft Excel and Minitab*. Upper Saddle River, NJ: Prentice Hall, 2001.

7. Microsoft Excel 2007. Redmond, WA: Microsoft Corporation, 2006.

Multiple Regression

11.1 The Multiple Regression Model

11.2 Coefficient of Multiple Determination

11.3 The Overall F Test

11.4 Residual Analysis for the Multiple Regression Model

11.5 Inferences Concerning the Population Regression Coefficients

One-Minute Summary

Test Yourself

Chapter 10 discussed the simple linear regression model that uses one numerical independent variable X to predict the value of a numerical dependent variable Y. Often you can make better predictions if you use more than one independent variable. This chapter introduces you to multiple regression models that use two or more independent variables (Xs) to predict the value of a dependent variable (Y).

11.1 The Multiple Regression Model

CONCEPT The statistical method that extends the simple linear regression model of Equation (10.1) on page 231 by assuming a straight-line or linear relationship between each independent variable and the dependent variable.

WORKED-OUT PROBLEM 1 In Chapter 10, when analyzing the moving company data, you used the cubic footage to be moved to predict the labor hours. In addition to cubic footage (X_1), now you are also going to consider the number of pieces of large furniture (such as beds, couches, china closets, and dressers) that need to be moved (X_2 = LARGE = number of pieces of large furniture to be moved).

Moving

Hours	Feet	Large	Hours	Feet	Large
24.00	545	3	25.00	557	2
13.50	400	2	45.00	1,028	5
26.25	562	2	29.00	793	4
25.00	540	2	21.00	523	3
9.00	220	1	22.00	564	3
20.00	344	3	16.50	312	2
22.00	569	2	37.00	757	3
11.25	340	1	32.00	600	3
50.00	900	6	34.00	796	3
12.00	285	1	25.00	577	3
38.75	865	4	31.00	500	4
40.00	831	4	24.00	695	3
19.50	344	3	40.00	1,054	4
18.00	360	2	27.00	486	3
28.00	750	3	18.00	442	2
27.00	650	2	62.50	1,249	5
21.00	415	2	53.75	995	6
15.00	275	2	79.50	1,397	7

Spreadsheet and calculator results for these data are as follows:

	A	B	C	D	E	F	G	H	I
1	Multiple Regression Analysis for Moving Company Study								
2									
3	*Regression Statistics*								
4	Multiple R	0.9658							
5	R Square	0.9327							
6	Adjusted R Square	0.9287							
7	Standard Error	3.9800							
8	Observations	36							
9									
10	ANOVA								
11		df	SS	MS	F	Significance F			
12	Regression	2	7248.71	3624.35	228.8049	4.55E-20			
13	Residual	33	522.73	15.84					
14	Total	35	7771.44						
15									
16		Coefficients	Standard Error	t Stat	P-value	Lower 95%	Upper 95%	Lower 95.0%	Upper 95.0%
17	Intercept	-3.9152	1.6738	-2.3391	0.0255	-7.3206	-0.5099	-7.3206	-0.5099
18	Cubic Feet Moved	0.0319	0.0046	6.9339	0.0000	0.0226	0.0413	0.0226	0.0413
19	Large	4.2228	0.9142	4.6192	0.0001	2.3629	6.0828	2.3629	6.0828

```
      DF  SS              ⋮    B0=-3.915221424 ⋮
  RG  2   7248.70560           CL COEFF ∕ T   P
  ER 33   522.731898           2 .0319243448
         F=228.8                    6.93    0.000
         P=0.000                3 4.222833883
       R-SQ=.9327                    4.62    0.000
       (ADJ).9287
       S=3.979995047
```

Net Regression Coefficients

CONCEPT   The coefficients that measure the change in Y per unit change in a particular X, holding constant the effect of the other X variables. (Net regression coefficients are also known as **partial regression coefficients**.)

INTERPRETATION   In the simple regression model in Chapter 10, the slope represents the change in Y per unit change in X and does not take into account any other variables besides the single independent variable included in the model. In the multiple regression model with two independent variables, there are two net regression coefficients: b_1, the slope of Y with X_1 represents the change in Y per unit change in X_1, taking into account the effect of X_2; and b_2, the slope of Y with X_2 represents the change in Y per unit change in X_2, taking into account the effect of X_1.

For example, the results for the moving company study show that the slope of Y with X_1, b_1 is 0.0319 and the slope of Y with X_2, b_2, is 4.2228. The slope b_1 means that for each increase of 1 unit in X_1, the value of Y is estimated to increase by 0.0319 units, holding constant the effect of X_2. In other words, holding constant the number of pieces of large furniture, for each increase of 1 cubic foot in the amount to be moved, the fitted model predicts that the labor hours are estimated to increase by 0.0319 hours.

The slope b_2 (+4.2228) means that for each increase of 1 unit in X_2, the value of Y is estimated to increase by 4.2228 units, holding constant the effect of X_1. In other words, holding constant the amount to be moved, for each additional piece of large furniture, the fitted model predicts that the labor hours are estimated to increase by 4.2228 hours.

Another way to interpret this "net effect" is to think of two moves with an equal number of pieces of large furniture. If the first move consists of 1 cubic foot more than the other move, the "net effect" of this difference is that the first move is predicted to take 0.0319 more labor hours than the other move. To interpret the net effect of the number of pieces of large furniture, you can consider two moves that have the same cubic footage. If the first move has one additional piece of large furniture, the net effect of this difference is that the first move is predicted to take 4.2228 more labor hours than the other move.

Adding the Y intercept, b_0, to the net regression coefficients b_1 and b_2 creates the multiple regression equation, which for a model with two independent variables is:

$$\text{Predicted } Y = b_0 + b_1 X_1 + b_2 X_2$$

For example, the results for the moving company study show that the Y intercept, b_0, is –3.915. This means that for this study the multiple regression equation is

Predicted value of labor hours = –3.915 + (0.0319 × cubic feet moved)

+ (4.2228 × large furniture)

(Recall from Chapter 10 that the Y intercept represents the estimated value of Y when X equals 0. In this example, because the cubic feet moved cannot be less than 0, the Y intercept has no practical interpretation. Recall, also, that a regression model is only valid within the ranges of the independent variables.)

Predicting the Dependent Variable *Y*

As in simple linear regression, you can use the multiple regression equation to predict values of the dependent variable.

WORKED-OUT PROBLEM 2   Using the multiple regression model developed in **WORKED-OUT PROBLEM 1**, you want to predict the labor hours for a move with 500 cubic feet with three large pieces of furniture to be moved. You predict that the labor hours for such a move are 24.715 (–3.915 + 0.0319 × 500 + 4.2228 × 3).

11.2 Coefficient of Multiple Determination

CONCEPT   The statistic that represents the proportion of the variation in Y that is explained by the set of independent variables included in the multiple regression model.

INTERPRETATION   The coefficient of multiple determination is analogous to the coefficient of determination (r^2) that measures the variation in Y that is explained by the independent variable X in the simple linear regression model (see Section 10.3).

WORKED-OUT PROBLEM 3   You need to calculate this coefficient for the moving company study. In the multiple regression results (see page 246), the ANOVA summary table shows that SSR is 7,248.71 and SST is 7,771.44. Therefore,

$$r^2 = \frac{\text{regression sum of squares}}{\text{total sum of squares}} = \frac{SSR}{SST}$$

$$= \frac{7{,}248.71}{7{,}771.44}$$

$$= 0.9327$$

The coefficient of multiple determination, ($r^2 = 0.9327$), indicates that 93.27% of the variation in labor hours is explained by the variation in the cubic footage and the variation in the number of pieces of large furniture to be moved.

11.3 The Overall *F* test

CONCEPT   The test for the significance of the overall multiple regression model.

INTERPRETATION   You use this test to determine whether a significant relationship exists between the dependent variable and the entire set of independent variables. Because there is more than one independent variable, you have the following null and alternative hypotheses:

> H_0: No linear relationship exists between the dependent variable and the independent variables.

> H_1: A linear relationship exists between the dependent variable and at least one of the independent variables.

The ANOVA summary table for the overall *F* test is as follows:

Source	Degrees of Freedom	Sum of Squares	Mean Square (Variance)	*F*
Regression	k	SSR	$MSR = \dfrac{SSR}{k}$	$F = \dfrac{MSR}{MSE}$
Error	$n - k - 1$	SSE	$MSE = \dfrac{SSE}{n - k - 1}$	
Total	$n - 1$	SST		

where

n = sample size

k = number of independent variables

WORKED-OUT PROBLEM 4   You need to perform this test for the moving company study. In the multiple regression results (see page 246), the ANOVA summary table shows that the F statistic is 228.849 and the p-value $= 0.000$ (shown in the results as 4.55E-20). Because the p-value $= 0.000 < 0.05$, you reject H_0 and conclude that at least one of the independent variables (cubic footage and/or the number of pieces of large furniture moved) is related to labor hours.

11.4 Residual Analysis for the Multiple Regression Model

In Section 10.5, you used residual analysis to evaluate whether the simple linear regression model was appropriate for a set of data. For the multiple regression model with two independent variables, you need to construct and analyze the following residual plots:

- Residuals versus the predicted value of Y
- Residuals versus the first independent variable X_1
- Residuals versus the second independent variable X_2
- Residuals versus time (if the data has been collected in time order)

The first residual plot examines the pattern of residuals versus the predicted values of Y. If the residuals show a pattern for different predicted values of Y, there is evidence of a possible curvilinear effect in at least one independent variable, a possible violation to the assumption of equal variance and/or the need to transform the Y variable. The second and third residual plots involve the independent variables. Patterns in the plot of the residuals versus an independent variable can indicate the existence of a curvilinear effect and, therefore, indicate the need to add a curvilinear independent variable to the multiple regression model (see References 1 and 2). The fourth plot is used to investigate patterns in the residuals to determine whether the independence assumption has been violated when the data are collected in time order.

WORKED-OUT PROBLEM 5   You need to perform a residual analysis for the multiple regression model for the moving company study. The residual plots for this model are shown on the following page.

In these plots, you see very little or no pattern in the relationship between the residuals and the predicted value of Y, the cubic feet moved (X_1), or the number of pieces of large furniture moved (X_2). You conclude that the multiple regression model is appropriate for predicting labor hours.

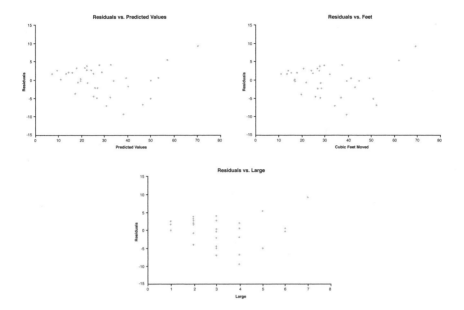

11.5 Inferences Concerning the Population Regression Coefficients

In Section 10.6, you tested the existence of the slope in a simple linear regression model to determine the significance of the relationship between X and Y. In addition, you constructed a confidence interval of the population slope. In this section, these procedures are extended to the multiple regression model.

Tests of Hypothesis

As with the simple linear regression model, you use a t test to test a hypothesis concerning the population slope: For a multiple regression model with two independent variables, the null hypothesis for each independent variable is that no linear relationship exists between labor hours and the independent variable holding constant the effect of the other independent variables. The alternative hypothesis is that a linear relationship exists between labor hours and the independent variable holding constant the effect of the other independent variables. The t test of significance for a particular regression coefficient is actually a test for the significance of adding a particular variable into a regression model given that the other variable is included.

WORKED-OUT PROBLEM 6   You need to test the hypothesis concerning the population slopes of the multiple regression model for the moving company study. From the results on page 246, for the cubic feet moved independent variable, the t statistic is 6.93 and the p-value is 0.000. Because the p-value is 0.000 < 0.05, you reject the null hypothesis and conclude that a linear relationship exists between labor hours and the cubic feet moved (X_1). For the number of pieces of large furniture moved independent variable, the t statistic is 4.62 and the p-value is 0.000. Because the p-value is 0.000 < 0.05, you reject the null hypothesis and conclude that a linear relationship exists between labor hours and the number of pieces of large furniture moved (X_2). You conclude that because each of the two independent variables is significant, both should be included in the regression model.

Confidence Interval Estimation

As you did in simple linear regression (see Section 10.6), you can construct confidence interval estimates of the slope. You calculate the confidence interval estimate of the population slope by multiplying the t statistic by the standard error of the slope and then adding and subtracting this product to the sample slope.

WORKED-OUT PROBLEM 7   You want to determine the confidence interval estimates of the slope from the moving company study multiple regression model. From the results on page 246, with 95% confidence, the lower limit for the slope of the number of feet moved with labor hours is 0.0226 hours and the upper limit is 0.0413 hours. The confidence interval indicates that for each increase of 1 cubic foot moved, labor hours are estimated to increase by at least 0.0226 hours but less than 0.0413 hours, holding constant the number of pieces of large furniture moved. With 95% confidence, the lower limit of the slope of the number of pieces of large furniture moved and labor hours is 2.3629 hours and the upper limit is 6.0828 hours. The confidence interval indicates that for each increase of one piece of large furniture moved, labor hours are estimated to increase by at least 2.3629 hours but less than 6.0827 hours holding constant the cubic footage moved.

calculator keys

Multiple Regression Using A2MULREG

First, enter the data values for your dependent variable Y and your independent X variables into matrix variable [**D**]. To do this:

- Press [**2nd**][x^{-1}][◀] and then press [**ENTER**] and select 4:[**D**] to display the MATRIX[D] screen.

- Type the number of rows, press [▶], type the number of columns, and press [▶]. (The number of rows will be the number of data values for each variable and the number of columns will be the number of independent variables plus 1.)

- In the table-like entry area that appears, enter the values for your dependent variable *Y* in the *first* column. Then enter the values for the independent variables in the subsequent columns. (Press [ENTER] after each entry.)

- When your data entry is completed, press [2nd] [MODE].

Next, use the Section AC3 instructions in Appendix A to load the **A2MULREG** program. When the program is loaded, press [PRGM] (to display the EXEC menu) and select the **A2MUL-REG** choice and press [ENTER]. The program begins and displays the prompt **prgmA2MULREG**. Press [ENTER] to display a second screen (which reminds you that data must be already entered into matrix variable [D]) and press [ENTER] a second time to display the **MULT REG+CORR** screen. Select **1:MULT REGRESSIO** and press [ENTER] to begin the multiple regression analysis. Continue with these steps:

- Type the number of independent *X* variables and press [ENTER] as the answer to **HOW MANY IND VAR?** (Type **2** and press [ENTER], if you are using the moving company study multiple regression data.)

- Always type **2** and press [ENTER] as the answer to **COL. OF VAR. 1?**.

- For each consecutive **COL OF VAR.** question, enter the proper column number. For example, type **3** and press [ENTER] as the answer to **COL. OF VAR. 2?**

After answering the COLumn OF (independent) VARiable questions, the program will display results similar to the first screen shown on page 247. Press [ENTER] to see results similar to the second screen shown on page 247. When you are finished viewing the results, press [ENTER] to display a screen labeled MAIN MENU.

From the MAIN MENU screen, you can calculate confidence interval estimates and prediction intervals (not discussed in this book) for the multiple regression, perform a residual analysis (see the next Calculator Keys), begin a new multiple regression model, or end the program. To end the program, select **4:QUIT** and press [ENTER].

calculator keys

Residual Analysis Using A2MULREG

From the MAIN MENU screen of the **A2MULREG** program (see preceding **Calculator Keys**), select **2:RESIDUALS** and press **[ENTER]** to display the RES OR STAND RES screen. (You may get the "Busy screen" while the calculator prepares for the analysis.) In this screen, select **1:RESIDUAL PLOT** and press **[ENTER]** to display the PLOT OF RESID screen. In this screen:

- Select **1:VS YHAT.** and press **[ENTER]** to display a plot of the residuals versus the predicted *Y* values.

- Select **2:VS AN IND VAR.** and answer the question **WHAT COL?** to display a plot of the residuals and a specific independent *X* variable.

To exit a plot screen, press **[ENTER]**, which returns you to the RES OR STAND RES screen from where you can continue with another plot or quit the program (by selecting **5:QUIT** and pressing **[ENTER]**).

A2MULREG residual analysis contains other features, such as being able to calculate the Durbin-Watson Statistic, which are beyond the scope of this book. More information about using **A2MULREG** can be found on the Texas Instruments Education Technology website, **education.ti.com**.

spreadsheet solution

Multiple Regression

Chapter 11 Multiple Regression ATP (shown on page 246) contains the regression results for the moving company study data as calculated by the Analysis ToolPak **Regression** procedure. This workbook also contains the residual plots that appear in Section 11.4.

Spreadsheet Tip ATT6 (in Appendix D) explains how to use the **Regression** procedure. As with simple linear regression, if you use OpenOffice.org Calc 3, you must use the **LINEST** function, discussed in Section E.4 of Appendix E to calculate regression results. (You can also use the **LINEST** function in Microsoft Excel versions 2003 or later.)

One-Minute Summary

Multiple regression

- Use several independent variables to predict a dependent variable
- Net regression coefficients
- Coefficient of Multiple Determination
- Overall F Test
- Residual analysis
- t test for the significance of each independent variable

Test Yourself

Short Answers

1. In a multiple regression model involving two independent variables, if b_1 is +3.0, it means that:
   - (a) The relationship between X_1 and Y is significant.
   - (b) The estimated value of Y increases by 3 units for each increase of 1 unit of X_1, holding X_2 constant.
   - (c) The estimated value of Y increases by 3 units for each increase of 1 unit of X_1, without regard to X_2.
   - (d) The estimated value of Y is 3 when X_1 equals zero.

2. The coefficient of multiple determination
   - (a) measures the variation around the predicted regression equation
   - (b) measures the proportion of variation in Y that is explained by X_1 and X_2
   - (c) measures the proportion of variation in Y that is explained by X_1 holding X_2 constant
   - (d) will have the same sign as b_1

3. In a multiple regression model, the value of the coefficient of multiple determination
   - (a) is between −1 and +1
   - (b) is between 0 and +1
   - (c) is between −1 and 0
   - (d) can be any number

Answer True or False:

4. The interpretation of the slope is different in a multiple linear regression

model as compared to a simple linear regression model.

5. The interpretation of the Y intercept is different in a multiple linear regression model as compared to a simple linear regression model.

6. In a multiple regression model with two independent variables, the coefficient of multiple determination measures the proportion of variation in Y that is explained by X_1 and X_2.

7. The slopes in a multiple regression model are called net regression coefficients.

8. The coefficient of multiple determination is calculated by taking the ratio of the regression sum of squares over the total sum of squares (SSR/SST) and subtracting that value from 1.

9. You have just developed a multiple regression model in which the value of coefficient of multiple determination is 0.35. To determine whether this indicates that the independent variables explain a significant portion of the variation in the dependent variable, you would perform an F-test.

10. From the coefficient of multiple determination, you cannot detect the strength of the relationship between Y and any individual independent variable.

Answers to Test Yourself Short Answers

1. b	6. True
2. b	7. True
3. b	8. False
4. True	9. True
5. False	10. True

Problems

1. The appraised value (in thousands of dollars), land area (in acres), and age of house (in years) was collected for a sample of 30 single-family homes located in Glen Cove, New York (see the following table). Develop a multiple linear regression model to predict appraised value based on land area of the property and age, in years.

GlenCove

Appraised Value	Land (acres)	Age
466.0	0.2297	46
364.0	0.2192	51
429.0	0.1630	29
548.4	0.4608	18

Appraised Value	Land (acres)	Age
405.9	0.2549	46
374.1	0.2290	88
315.0	0.1808	48
749.7	0.5015	7
217.7	0.2229	52
635.7	0.1300	15
350.7	0.1763	54
455.0	0.4200	48
356.2	0.2520	46
271.7	0.1148	12
304.3	0.1693	64
288.4	0.1714	52
396.7	0.3849	44
613.5	0.6545	46
314.1	0.1722	52
363.5	0.1435	78
364.3	0.2755	71
305.1	0.1148	97
441.7	0.3636	45
353.1	0.1474	41
463.3	0.2281	40
320.0	0.4626	82
332.8	0.1889	54
276.6	0.1228	44
397.0	0.1492	34
221.9	0.0852	94

(a) State the multiple regression equation.
(b) Interpret the meaning of the slopes, b_1 and b_2, in this problem.
(c) Explain why the regression coefficient, b_0, has no practical meaning in the context of this problem.

(d) Predict the appraised value for a house that has a land area of 0.25 acres and is 45 years old.

(e) Compute the coefficient of multiple determination, r^2, and interpret its meaning.

(f) Perform a residual analysis on the results and determine the adequacy of the model.

(g) Determine whether a significant relationship exists between the appraised value and the two independent variables (land area of a property and age of a house) at the 0.05 level of significance.

(h) At the 0.05 level of significance, determine whether each independent variable makes a significant contribution to the regression model. On the basis of these results, indicate the independent variables to include in this model.

(i) Construct a 95% confidence interval estimate of the population slope between appraised value and land area of a property and between appraised value and age.

2. Measuring the height of a California redwood tree is a very difficult undertaking because these trees grow to heights of more than 300 feet. People familiar with these trees understand that the height of a California redwood tree is related to other characteristics of the tree, including the diameter of the tree at the breast height of a person and the thickness of the bark of the tree. The following data represent the height (in feet), diameter at breast height of a person, and bark thickness for a sample of 21 California redwood trees:

Redwood

Height	Diameter at breast height	Bark thickness
122.0	20	1.1
193.5	36	2.8
166.5	18	2.0
82.0	10	1.2
133.5	21	2.0
156.0	29	1.4
172.5	51	1.8
81.0	11	1.1
148.0	26	2.5
113.0	12	1.5
84.0	13	1.4
164.0	40	2.3

Height	Diameter at breast height	Bark thickness
203.3	52	2.0
174.0	30	2.5
159.0	22	3.0
205.0	42	2.6
223.5	45	4.3
195.0	54	4.0
232.5	39	2.2
190.5	36	3.5
100.0	8	1.4

(a) State the multiple regression equation that predicts the height of a tree based on the tree's diameter at breast height and the thickness of the bark.

(b) Interpret the meaning of the slopes in this equation.

(c) Predict the height for a tree that has a breast diameter of 25 inches and a bark thickness of 2 inches.

(d) Interpret the meaning of the coefficient of multiple determination in this problem.

(e) Perform a residual analysis on the results and determine the adequacy of the model.

(f) Determine whether a significant relationship exists between the height of redwood trees and the two independent variables (breast diameter and the bark thickness) at the 0.05 level of significance.

(g) At the 0.05 level of significance, determine whether each independent variable makes a significant contribution to the regression model. Indicate the independent variables to include in this model.

(h) Construct a 95% confidence interval estimate of the population slope between the height of the redwood trees and breast diameter and between the height of redwood trees and the bark thickness.

3. Professional basketball has truly become a sport that generates interest among fans around the world. More and more players come from outside the United States to play in the National Basketball Association (NBA). You want to develop a regression model to predict the number of wins achieved by each NBA team, based on field goal (shots made) percentage for the team and for the opponent, using the following data:

NBA2008

Team	Wins	Field Goal %	Field Goal % Allowed
Atlanta	37	0.454	0.463
Boston	66	0.475	0.419
Charlotte	32	0.452	0.466
Chicago	33	0.435	0.453
Cleveland	45	0.439	0.455
Dallas	51	0.464	0.443
Denver	50	0.470	0.457
Detroit	59	0.458	0.437
Golden State	48	0.459	0.468
Houston	55	0.448	0.433
Indiana	36	0.444	0.454
L.A. Clippers	23	0.438	0.467
L.A. Lakers	57	0.476	0.445
Memphis	22	0.454	0.480
Miami	15	0.443	0.468
Milwaukee	26	0.449	0.480
Minnesota	22	0.451	0.472
New Jersey	34	0.443	0.456
New Orleans	56	0.466	0.460
New York	23	0.439	0.474
Orlando	52	0.474	0.446
Philadelphia	40	0.460	0.461
Phoenix	55	0.500	0.456
Portland	41	0.448	0.451
Sacramento	38	0.464	0.466
San Antonio	56	0.457	0.444
Seattle	20	0.444	0.461
Toronto	41	0.468	0.458
Utah	54	0.497	0.461
Washington	43	0.446	0.461

(a) State the multiple regression equation.

(b) Interpret the meaning of the slopes in this equation.

(c) Predict the number of wins for a team that has a field goal percentage of 45% and an opponent field goal percentage of 44%.

(d) Perform a residual analysis on your results and determine the adequacy of the fit of the model.

(e) Is there a significant relationship between number of wins and the two independent variables (field goal percentage for the team and for the opponent) at the 0.05 level of significance?

(f) Interpret the meaning of the coefficient of multiple determination in this problem.

(g) At the 0.05 level of significance, determine whether each independent variable makes a significant contribution to the regression model. Indicate the most appropriate regression model for this set of data.

(h) Construct a 95% confidence interval estimate of the population slope between wins and field goal percentage for the team and between wins and field goal percentage for the opponent.

Auto

4. A consumer organization wants to develop a regression model to predict gasoline mileage (as measured by miles per gallon) based on the horsepower of the car's engine and the weight of the car (in pounds). A sample of 50 recent car models was selected (see the Auto file).

(a) State the multiple regression equation.

(b) Interpret the meaning of the slopes, b_1 and b_2, in this problem.

(c) Explain why the regression coefficient, b_0, has no practical meaning in the context of this problem.

(d) Predict the miles per gallon for cars that have 60 horsepower and weigh 2,000 pounds.

(e) Perform a residual analysis on your results and determine the adequacy of the fit of the model.

(f) Compute the coefficient of multiple determination, r^2, and interpret its meaning.

(g) Determine whether a significant relationship exists between gasoline mileage and the two independent variables (horsepower and weight) at the 0.05 level of significance.

(h) At the 0.05 level of significance, determine whether each independent variable makes a significant contribution to the regression model. On the basis of these results, indicate the independent variables to include in this model.

(i) Construct a 95% confidence interval estimate of the population slope between gasoline mileage and horsepower and between gasoline mileage and weight.

Answers to Problems

1.   (a) Predicted appraised value = 400.8057 + 456.4485 land – 2.4708 age

    (b) For a given age, each increase by one acre in land area is estimated to result in an increase in appraised value by $456.45 thousands. For a given acreage, each increase of one year in age is estimated to result in the decrease in appraised value by $2.47 thousands.

    (c) The interpretation of b_0 has no practical meaning here because it would represent the estimated appraised value of a new house that has no land area.

    (d) Predicted appraised value = 400.8057 + 456.4485(0.25) – 2.4708(45) = $403.73 thousands.

    (e) $r^2 = 0.5813$; so, 58.13% of the variation in appraised value of a house can be explained by variation in land area and age of the house.

    (f) There is no particular pattern in the residual plots, and the model appears to be adequate.

    (g) $F = 18.7391$; p-value is virtually zero. Reject H_0 at 5% level of significance. Evidence exists of a significant linear relationship between appraised value and the two explanatory variables.

    (h) For land area: $t = 4.0922 > 2.0518$ or p-value = 0.00035 < 0.05 Reject H_0. Land area makes a significant contribution to the regression model after age is included. For age: $t = -3.6295 < -2.0518$ or p-value = 0.0012 < 0.05 Reject H_0. Age makes a significant contribution to the regression model after land area is included. Therefore, both land area and age should be included in the model.

    (i) $227.5865 < \beta_1 < 685.3104$ $-3.8676 < \beta_2 < -1.074$

2.   (a) Predicted height = 62.1411 + 2.0567 diameter of the tree at breast height of a person (in inches) + 15.6418 thickness of the bark (in inches).

    (b) Holding constant the effects of the thickness of the bark, for each additional inch in the diameter of the tree at breast height of a person, the height of the tree is estimated to increase by 2.0567 feet. Holding constant the effects of the diameter of the tree at breast height of a person, for each additional inch in the thickness of the bark, the height of the tree is estimated to increase by 15.6418 feet.

    (c) Predicted height = 62.1411 + 2.0567 (25) + 15.6418 (2) = 144.84 feet

    (d) $r^2 = 0.7858$. 78.58% of the total variation in the height of the tree can be explained by the variation in the diameter of the tree at breast height of a person and the thickness of the bark of the tree.

(e) The plot of the residuals against bark thickness indicates a potential pattern that might require the addition of curvilinear terms. One value appears to be an outlier in both plots.

(f) $F = 33.0134$ with 2 and 18 degrees of freedom. p-value = virtually $0 < 0.05$. Reject H_0. At least one of the independent variables is linearly related to the dependent variable.

(g) Breast height diameter: $t = 4.6448 > 2.1009$ or p-value = 0.0002 < 0.05 Reject H_0. Breast height diameter makes a significant contribution to the regression model after bark thickness is included. Bark thickness: $t = 2.1882 > 2.1009$ or p-value = 0.0421 < 0.05 Reject H_0. Bark thickness makes a significant contribution to the regression model after breast height diameter is included. Therefore, both breast height diameter and bark thickness should be included in the model.

(h) $1.1264 \le \beta_1 \le 2.9870$ $0.6238 \le \beta_2 \le 30.6598$

3. (a) Predicted wins = 153.307 + 406.800 field goal % in decimals − 652.4386 opponent field goal % in decimals

(b) For a given opponent field goal %, each increase of 1% in field goal % increases the estimated number of wins by 4.0678. For a given field goal %, each increase of 1% in opponent field goal % decreases the estimated number of wins by 6.5244.

(c) Predicted wins = 153.3070 + 406.8000 (0.45) − 652.4386 (0.44) = 49.2849

(d) There is no particular pattern in the Field Goal % or the Field Goal % Allowed residual plots.

(e) $F = MSR/MSE = 2{,}196.9343/42.5975 = 51.5743$; p-value = 0.0000 < 0.05. Reject H_0 at 5% level of significance. Evidence of a significant linear relationship exists between number of wins and the two explanatory variables.

(f) $r^2 = SSR/SST = 4{,}393.8686/5{,}544.0 = 0.7925$. 79.25% of the variation in number of wins can be explained by variation in field goal % and opponent field goal %.

(g) For X_1: $t_{STAT} = b_1 / S_{b_1} = 5.2248 > 2.0518$ and p-value 0.0000 < 0.05, reject H_0. There is evidence that the variable X_1 contributes to a model already containing X_2. For X_2: $t_{STAT} = b_2 / S_{b2} = -7.0208$ < −2.0518 and p-value 0.0000 < 0.05, reject H_0. Both variables X_1 and X_2 should be included in the model.

(h) $247.0347 < \beta_1 < 566.5252$ $-843.114 < \beta_2 < -461.7633$

4. (a) Predicted miles per gallon (MPG) = 58.15708 − 0.11753 horsepower − 0.00687 weight

(b) For a given weight, each increase of one unit in horsepower is estimated to result in a decrease in MPG of 0.11753. For a given horsepower, each increase of one pound is estimated to result in the decrease in MPG of 0.00687.

(c) The interpretation of b_0 has no practical meaning here because it would have involved estimating gasoline mileage when a car has 0 horsepower and 0 weight.

(d) Predicted miles per gallon (MPG) = 58.15708 – 0.11753 (60) – 0.00687 (2,000) = 37.365 MPG

(e) There appears to be a curvilinear relationship in the plot of the residuals against the predicted values of MPG and the horsepower. Thus, curvilinear terms for the independent variables should be considered for inclusion in the model.

(f) F = 1,225.98685/17.444 = 70.2813; p-value = 0.0000 < 0.05 Reject H_0. Evidence of a significant linear relationship exists between MPG and horsepower and/or weight.

(g) r^2 = 2,451.974/3,271.842 = 0.7494 74.94% of the variation in MPG can be explained by variation in horsepower and variation in weight.

(h) For horsepower: $t < -3.60 < -2.0117$ or p-value = 0.0008 < 0.05 Reject H_0. There is evidence that horsepower contributes to a model already containing weight. For weight: $t < -4.9035 < -2.0117$ or p-value = 0.0000 < 0.05 Reject H_0. There is evidence that weight contributes to a model already containing horsepower. Both variables X_1 and X_2 should be included in the model.

(i) $-0.1832 < \beta_1 < -0.0519$ $-0.0097 < \beta_2 < -0.00405$

References

1. Berenson, M. L., D. M. Levine, and T. C. Krehbiel. *Basic Business Statistics: Concepts and Applications*, 11th Ed. (Upper Saddle River, NJ: Prentice Hall, 2009).

2. Kutner, M. H., C. Nachtsheim, J. Neter, and W. Li. *Applied Linear Statistical Models*, 5th Ed. (New York: McGraw-Hill-Irwin, 2005).

3. Microsoft Excel 2007. Redmond, WA: Microsoft Corp., 2006.

CHAPTER

12

Quality and Six Sigma Applications of Statistics

12.1 Total Quality Management
12.2 Six Sigma
12.3 Control Charts
12.4 The p Chart
12.5 The Parable of the Red Bead Experiment:
    Understanding Process Variability
12.6 Variables Control Charts for the Mean and Range
    Important Equations
    One-Minute Summary
    Test Yourself

In recent times, improving quality and productivity have become essential goals for all organizations. However, monitoring and measuring such improvements can be problematic if subjective judgments about quality are made. A set of statistical techniques known as statistical process control as well as management practices such as Total Quality Management and Six Sigma assist in quality improvement by using statistical methods to measure sources of variation and reduce defects.

12.1 Total Quality Management

During the past 30 years, the renewed interest in quality and productivity in the United States followed as a reaction to perceived improvements of Japanese industry that had begun as early as 1950. Individuals such as W. Edwards Deming, Joseph Juran, and Kaoru Ishikawa developed an approach that focused on continuous improvement of products and services through an increased emphasis on statistics, process improvement, and optimization of the total system. This approach, widely known as **total quality management (TQM)**, is characterized by these themes:

important point

- The primary focus is on process improvement.
- Most of the variation in a process is due to the system and not the individual.
- Teamwork is an integral part of a quality management organization.
- Customer satisfaction is a primary organizational goal.
- Organizational transformation must occur in order to implement quality management.
- Fear must be removed from organizations.
- Higher quality costs less, not more, but requires an investment in training.

As this approach became more familiar, the federal government of the United States began efforts to encourage increased quality in American business, starting, for example, the annual competition for the Malcolm Baldrige Award, given to companies making the greatest strides in improving quality and customer satisfaction with their products and services. W. Edwards Deming became a more prominent consultant and widely discussed his "14 points for management."

1. Create constancy of purpose for improvement of product and service.
2. Adopt the new philosophy.
3. Cease dependence on inspection to achieve quality.
4. End the practice of awarding business on the basis of price tag alone. Instead, minimize total cost by working with a single supplier.
5. Improve constantly and forever every process for planning, production, and service.
6. Institute training on the job.
7. Adopt and institute leadership.
8. Drive out fear.
9. Break down barriers between staff areas.
10. Eliminate slogans, exhortations, and targets for the workforce.
11. Eliminate numerical quotas for the workforce and numerical goals for management.
12. Remove barriers that rob people of pride of workmanship. Eliminate the annual rating or merit system.
13. Institute a vigorous program of education and self-improvement for everyone.
14. Put everyone in the company to work to accomplish the transformation.

Although Deming's points were thought-provoking, some criticized his approach for lacking a formal, objective accountability. Many managers of large-scale organizations, used to seeing economic analyses of policy changes, needed a more prescriptive approach.

12.2 Six Sigma

One methodology, inspired by earlier TQM efforts, that attempts to apply quality improvement with increased accountability is the Six Sigma approach, originally conceived by Motorola in the mid-1980s. Refined and enhanced over the years, and famously applied at other large firms such as General Electric, Six Sigma was developed as a way to cut costs while improving efficiency. As with earlier total quality management approaches, Six Sigma relies on statistical process control methods to find and eliminate defects and reduce product variation.

CONCEPT   The quality management approach that is designed to create processes that result in no more than 3.4 defects per million.

INTERPRETATION   Six Sigma considers the variation of a process. Recall from Chapter 3, that the lowercase Greek letter sigma (σ) represents the population standard deviation, and recall from Chapter 5 that the range -6σ to $+6\sigma$ in a normal distribution includes virtually all (specifically, 0.999999998) of the probability or area under the curve. The Six Sigma approach assumes that the process might shift as much as 1.5 standard deviations over the long term. Six standard deviations minus a 1.5 standard deviation shift produces a 4.5 standard deviation goal. The area under the normal curve outside 4.5 standard deviations is approximately 3.4 out of a million (0.0000034).

The Six Sigma DMAIC Model

Unlike other quality management approaches, Six Sigma seeks to help managers achieve measurable, bottom-line results in a relatively short three to six-month period of time. This has enabled Six Sigma to garner strong support from top management of many companies (see References 1, 3, 7, 8, and 13).

To guide managers in their task of affecting short-term results, Six Sigma uses a five-step approach known as the **DMAIC** model, for the names of steps in the process: Define, Measure, Analyze, Improve, and Control. This model can be summarized as follows:

- **Define**—The problem to be solved is defined along with the costs, benefits of the project, and the impact on the customer.

- **Measure**—Operational definitions for each critical-to-quality (CTQ) characteristic must be developed. In addition, the measurement procedure must be verified so that it is consistent over repeated measurements. Finally, baseline data is collected to determine the capability and stability of the current process.

- **Analyze**—The root causes of why defects can occur need to be determined along with the variables in the process that cause these defects to occur. Data are collected to determine the underlying value for each process variable often using control charts (discussed in Sections 12.3 through 12.6).

- **Improve**—The importance of each process variable on the Critical-To-Quality (CTQ) characteristic are studied using designed experiments. The objective is to determine the best level for each variable that can be maintained in the long term.

- **Control**—This phase of a Six Sigma project focuses on the maintenance of improvements that have been made in the Improve phase. A risk abatement plan is developed to identify elements that can cause damage to a process.

Implementation of the Six Sigma approach requires intensive training in the DMAIC model as well as a data-oriented approach to analysis that uses designed experiments and various statistical methods, such as the control chart methods discussed in the remainder of the chapter.

12.3   Control Charts

CONCEPT   Control charts monitor variation in a characteristic of a product or service by focusing on the variation in a process over time.

INTERPRETATION   Control charts aid in quality improvement by letting you assess the stability and capability of a process. Control charts attempt to help you separate special or assignable causes of variation (defined next) from chance or common causes of variation. Control charts are divided into two types called attribute control charts and variables control charts.

- **Attribute control charts**, such as the *p* chart discussed in Section 12.4, are used to evaluate categorical data. If you wanted to study the proportion of newspaper ads that have errors or the proportion of trains that are late, you would use attribute control charts.

- **Variables control charts** are used for continuous data. If you wanted to study the waiting time at a bank or the weight of packages of candy, you would use variables control charts. Variables control charts contain more information than attribute charts and are generally used in pairs, such as the range chart and the mean chart.

Special or Assignable Causes of Variation

CONCEPT   Variation that represents large fluctuations or patterns in the data that is not inherent to a process.

EXAMPLE   If during your process of getting ready to go to work or school there is a leak in a toilet that needs immediate attention, your time to get ready will certainly be affected. This is special cause variation, because it is not a cause of variation that can be expected to occur every day, and there-fore it is not part of your everyday process of getting ready (at least you hope it is not!).

important point

INTERPRETATION   Special cause variation is the variation that is not always present in every process. It is variation that occurs for special reasons that usually can be explained.

Chance or Common Causes of Variation

CONCEPT   Variation that represents the inherent variability that exists in a process over time. This consists of the numerous small causes of variability that operate randomly or by chance.

EXAMPLE   Your process of getting ready to go to work or school has com-mon cause variation, because small variations exist in how long it takes you to perform the activities, such as making breakfast and getting dressed, that are part of your get-ready process from day to day.

INTERPRETATION   Common cause variation is the variation that is always present in every process. Typically, this variation can be reduced only by changing the process itself.

important point

Distinguishing between these two causes of variation is crucial, because only special causes of variation are not considered part of a process and therefore are correctable, or exploitable, without changing the system. Common causes of variation occur randomly or by chance and can be reduced only by chang-ing the system.

Control charts enable you to monitor the process and determine the presence of special causes. Control charts help prevent two types of errors. The first type of error involves the belief that an observed value represents special cause variation when actually the error is due to the common cause variation of the system. An example of this type of error occurs if you were to single out someone for disciplinary action based on having more errors than anyone else when actually the variation in errors was due only to the common cause variation in the system. Treating common causes of variation as special cause variation can result in overadjustment of a process that results in an accom-panying increase in variation.

The second type of error involves treating special cause variation as if it is common cause variation and not taking immediate corrective action when it is necessary. An example of this type of error occurs if you did not single someone out for disciplinary action based on having more errors than anyone else when the large number of errors made by the person could be explained and subsequently corrected. Although these errors can still occur when a control chart is used, they are far less likely.

Control Limits

CONCEPT   The upper and lower boundaries of the expected variation from the mean of the process under study. These boundaries are within ±3 standard deviations of the mean of the process. The value that is +3 standard deviations above the process mean is called the **upper control limit** (*UCL*). The value that is −3 standard deviations below the process mean is called the **lower control limit** (*LCL*). (If the value that is −3 standard deviations is less than zero, the lower control limit is set to zero.)

INTERPRETATION   Control limits are plotted with a center line that represents the mean of the process. Depending on the control chart being used, the process mean could be the mean proportion, the mean of a group of means, or the mean of the ranges.

important point

You examine a control chart to find whether any pattern exists in the values over time and determining whether any points fall outside the control limits. The simplest rule for detecting the presence of a special cause is one or more points falling beyond the ±3 standard deviation limits of the chart. The chart can be made more sensitive and effective in detecting out-of-control points if other signals and patterns that are unlikely to occur by chance alone are considered.

Two other simple rules enable you to detect a shift in the mean level of a process:

- Eight or more consecutive points lie above the center line, or eight or more consecutive points lie below the center line.
- Eight or more consecutive points move upward in value, or eight or more consecutive points move downward in value.

12.4 **The *p* Chart**

The most common attribute chart is the *p* chart, which is used for categorical variables.

CONCEPT   The control chart used to study a process that involves the proportion of items with a characteristic of interest, such as the proportion of newspaper ads with errors. Sample sizes in a *p* chart can remain constant or can vary.

INTERPRETATION   In the *p* chart, the process mean is the mean proportion of nonconformances. The mean proportion is computed from

$$\text{mean proportion} = \frac{\text{total number of nonconformances}}{\text{total number in all samples}}$$

To calculate the control limits, the mean sample size first needs to be calculated:

$$\text{mean sample size} = \frac{\text{total number of samples}}{\text{number of groups}}$$

Upper control limit (*UCL*) =

$$\text{mean proportion} + 3\sqrt{\frac{(\text{mean proportion})(1 - \text{mean proportion})}{\text{mean sample size}}}$$

Lower control limit (*LCL*) =

$$\text{mean proportion} - 3\sqrt{\frac{(\text{mean proportion})(1 - \text{mean proportion})}{\text{mean sample size}}}$$

To use a *p* chart, the following three statements must be true:

- There are only two possible outcomes for an event. An item must be found to be either conforming or nonconforming.
- The probability, *p*, of a nonconforming item is constant.
- Successive items are independent.

WORKED-OUT PROBLEM 1   You are part of a team in an advertising production department of a newspaper that is trying to reduce the number and dollar amount of the advertising errors. You collect data that tracks the number of ads with errors on a daily basis, excluding Sundays (which is considered to be substantially different from the other days). Data relating to the number of ads with errors in the last month are shown in the following table.

AdErrors

Day	Number of Ads with Errors	Number of Ads
1	4	228
2	6	273
3	5	239
4	3	197
5	6	259
6	7	203
7	8	289
8	14	241
9	9	263
10	5	199
11	6	275
12	4	212
13	3	207
14	5	245
15	7	266
16	2	197
17	4	228
18	5	236
19	4	208
20	3	214
21	8	258
22	10	267
23	4	217
24	9	277
25	7	258

You can use a p chart for these data because each ad is classified as having errors or not having errors, the probability of an ad with an error is assumed to be constant from day to day, and each ad is considered independent of the other ads.

A *p* chart prepared in Microsoft Excel for the newspaper ads data is shown in the following figure:

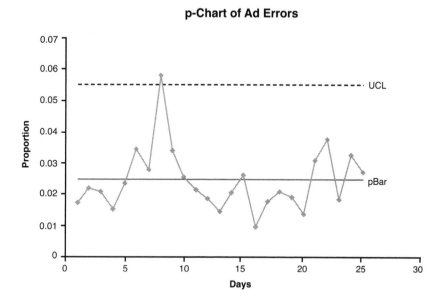

p-Chart of Ad Errors

For these data, the total number of nonconformances is 148, the total number in all samples is 5,956, and the number of groups is 25. Using these values, the mean proportion is 0.0248, and the mean sample size is 238.24, as shown:

$$\text{mean proportion} = \frac{\text{total number of nonconformances}}{\text{total number in all samples}}$$

$$= \frac{148}{5,956} = 0.0248$$

$$\text{mean sample size} = \frac{\text{total number of samples}}{\text{number of groups}}$$

$$= \frac{5,956}{25} = 238.24$$

Using the mean proportion and mean sample size values, the *UCL* is 0.0551 and the *LCL* is 0, as shown:

Upper control limit (*UCL*) =

$$0.0248 + 3\sqrt{\frac{(\text{mean proportion})(1 - \text{mean proportion})}{\text{mean sample size}}}$$

$$0.0248 + 3\sqrt{\frac{(0.0248)(1 - 0.0248)}{238.24}}$$

$$UCL = 0.0248 + 0.0303 = 0.0551$$

Lower control limit (LCL) =

$$0.0248 - 3\sqrt{\frac{(\text{mean proportion})(1 - \text{mean proportion})}{\text{mean sample size}}}$$

$$0.0248 - 3\sqrt{\frac{(0.0248)(1 - 0.0248)}{238.24}}$$

$$LCL = 0.0248 - 0.0303 = -0.0054$$

Because the calculated value is less than 0, the *LCL* is set at 0.

Using the rules for determining out-of-control points, you observe that point 8 is above the upper control limit. None of the other rules seems to be violated. There are no instances when eight consecutive points move upward or downward, nor are there eight consecutive points on one side of the center line.

Upon further investigation, you learn that point 8 corresponds to the day there was an employee from another work area assigned to the processing of the ads, because several employees were out ill. Your group brainstorms ways of avoiding such a problem in the future and recommends that a team of people from other work areas receive training on the work done by this area.

equation
blackboard
(optional)

You use these symbols to write the equations for the lower and upper control limits of a *p* chart:

- A subscripted uppercase italic X, X_i, for the number of nonconforming items in a group

- A subscripted lowercase italic n, n_i, for the sample or subgroup size for a group

- A lowercase italic k, k, for the number of groups selected

- A lowercase italic n bar, $\bar{n}$, for the mean group size

- A subscripted lowercase italic p, p_i, for the proportion of nonconforming items for a group
- A lowercase italic p bar, $\bar{p}$, for the mean proportion of nonconforming items

You first define $\bar{p}$ and $\bar{n}$ as

$$\bar{p} = \frac{\sum_{i=1}^{k} X_i}{\sum_{i=1}^{k} n_i} = \frac{\text{total number of nonconformances}}{\text{total sample size}}$$

and

$$\bar{n} = \frac{\sum_{i=1}^{k} n_i}{k} = \frac{\text{total sample size}}{\text{number of groups}}$$

You then use these symbols to write the equations for the control limits as follows:

$$LCL = \bar{p} - 3\sqrt{\frac{\bar{p}(1-\bar{p})}{n}}$$

$$UCL = \bar{p} + 3\sqrt{\frac{\bar{p}(1-\bar{p})}{\bar{n}}}$$

You also use some of the symbols to define the proportion of nonconforming items for group i as follows:

$$p_i = \frac{X_i}{n_i}$$

For the advertising errors data, the number of ads with errors is 148, the total sample size is 5,956, and there are 25 groups. Thus,

$$\bar{p} = \frac{148}{5,956} = 0.0248$$

and

$$\bar{n} = \frac{\sum_{i=1}^{k} n_i}{k} = \frac{\text{total sample size}}{\text{number of groups}} = \frac{5,956}{25} = 238.24$$

(*continues*)

so that

$$0.0248 + 3\sqrt{\frac{(0.0248)(1 - 0.0248)}{238.24}}$$

$$UCL = 0.0248 + 0.0303 = 0.0551$$

$$0.0248 - 3\sqrt{\frac{(0.0248)(1 - 0.0248)}{238.24}}$$

$$LCL = 0.0248 - 0.0303 = -0.0054$$

Because the LCL is less than 0, it is set at 0.

12.5 The Parable of the Red Bead Experiment: Understanding Process Variability

Imagine that you have been selected to appear on a new reality television series about job performance excellence. Over several days, you are assigned different tasks and your results are compared with your competitors.

The current task involves visiting the W.E. Beads Company and helping to select groups of 50 white beads for sale from a pool of 4,000 beads. You are told that W.E. Beads regularly tries to produce and sell only white beads, but that an occasional red bead gets produced in error. Unknown to you, the producer of the series has arranged that the pool of 4,000 beads contains 800 red beads to see how you and the other participants will react to this special challenge.

To select the groups of 50 beads, you and your competitors will be sharing a special scoop that can extract exactly 50 beads in one motion. You are told to hold the scoop at exactly an angle of 41 degrees to the vertical and that you will have three turns, simulating three days of production. At the end of each "day," two judges will independently count the number of red beads you select with the scoop and report their findings to a chief judge who might give out an award for exceptional job performance. To make things fair, after a group of 50 beads has been extracted, they will be returned to the pool so that every participant will always be selecting from the same pool of 4,000.

At the end of the three days, the judges, plus two famous business executives, will meet in a management council to discuss which worker deserves a promotion to the next task and which worker should be sent home from the competition. The competition results are as follows:

Contestant	Day 1	Day 2	Day 3	All 3 Days
You	9 (18%)	11 (22%)	6 (12%)	26 (17.33%)
A	12 (24%)	12 (24%)	8 (16%)	32 (21.33%)
B	13 (26%)	6 (12%)	12 (24%)	31 (20.67%)
C	7 (14%)	9 (18%)	8 (16%)	24 (16.0%)
All four workers:	41	38	34	113
Mean ($\overline{X}$)	10.25	9.5	8.5	9.42
Proportion	20.5%	19%	17%	18.83%

From the results, you observe several phenomena. On each day, some of the workers were above the mean and some below the mean. On day 1, C did best; but on day 2, B (who had the worst record on day 1) was best; and on day 3, you were the best. You are hopeful that your great job performance on day 3 will attract notice; if the decisions are solely based on job performance, however, whom would you promote and whom would you fire?

Deming's Red Bead Experiment

The description of the reality series is very similar to a famous demonstration that has become known as the **red bead experiment** that the statistician W. Edwards Deming performed during many lectures. In both the experiment and the imagined reality series, the workers have very little control over their production, even though common management practice might imply otherwise, and there are way too many managers officiating. Among the points about the experiment that Deming would make during his lectures are these:

important point

- Variation is an inherent part of any process.
- Workers work within a system over which they have little control. It is the system that primarily determines their performance.
- Only management can change the system.
- Some workers will always be above the mean, and some workers will always be below the mean.

How then can you explain all the variation? A *p* chart of the data puts the numbers into perspective and reveals that all the values are within the control limits, and no patterns exist in the results (see the following chart). The differences between you and the other participants merely represent the common cause variation expected in a stable process.

Red Bead Experiment

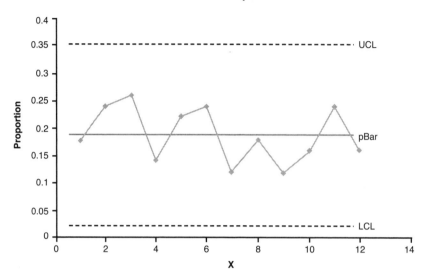

12.6 Variables Control Charts for the Mean and Range

Variables control charts can be used to monitor a process for a numerical variable such as bank waiting time. Because numerical variables provide more information than the proportion of nonconforming items, these charts are more sensitive in detecting special cause variation than the p chart. Variables charts are typically used in pairs. One chart monitors the variation in a process, while the other monitors the process mean. The chart that monitors variability must be examined first, because if it indicates the presence of out-of-control conditions, the interpretation of the chart for the mean will be misleading.

One of the most commonly employed pair of charts is the $\overline{X}$ chart used in conjunction with the R chart. The group range, R, is plotted on the R chart, which monitors process variability. The group mean is plotted on the $\overline{X}$ chart, which monitors the central tendency of the process.

WORKED-OUT PROBLEM 2   You want to study waiting times of customers for teller service at a bank during the peak 12 noon to 1 p.m. lunch hour. You select a group of four customers (one at each 15-minute interval during the hour) and measure the time in minutes from the point each customer enters the line to when he or she begins to be served. The results over a four-week period are as follows:

BankTime

Day	Time in Minutes			
1	7.2	8.4	7.9	4.9
2	5.6	8.7	3.3	4.2
3	5.5	7.3	3.2	6.0
4	4.4	8.0	5.4	7.4
5	9.7	4.6	4.8	5.8
6	8.3	8.9	9.1	6.2
7	4.7	6.6	5.3	5.8
8	8.8	5.5	8.4	6.9
9	5.7	4.7	4.1	4.6
10	3.7	4.0	3.0	5.2
11	2.6	3.9	5.2	4.8
12	4.6	2.7	6.3	3.4
13	4.9	6.2	7.8	8.7
14	7.1	6.3	8.2	5.5
15	7.1	5.8	6.9	7.0
16	6.7	6.9	7.0	9.4
17	5.5	6.3	3.2	4.9
18	4.9	5.1	3.2	7.6
19	7.2	8.0	4.1	5.9
20	6.1	3.4	7.2	5.9

R and $\bar{X}$ charts for these data appear on the next page:

Reviewing the R chart, you can see that none of the points on the R chart are outside of the control limits, and there are no other signals indicating a lack of control. This suggests that no special sources of variation are present.

Reviewing the $\bar{X}$ chart, you can see that none of the points on the $\bar{X}$ chart are outside of the control limits, and there are no other signals indicating a lack of control. This also suggests that no special sources of variation are

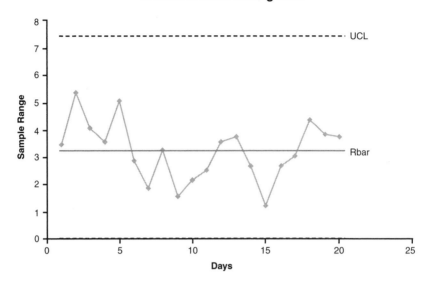

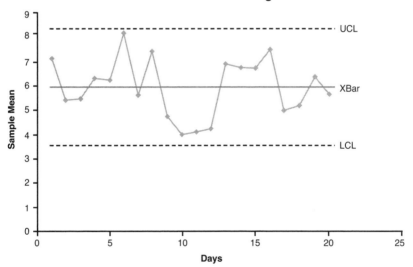

present. If management wants to reduce the variation in the waiting times or lower the mean waiting time, you conclude that changes in the process need to be made.

equation blackboard (optional)

interested in math?

Equations for the Lower and Upper Control Limits for the Range

You use the following symbols to write the equations for the lower and upper control limits for the range:

- A subscripted uppercase italic R, R_i, for the range of n observations at time i
- A lowercase italic k, k, for the number of groups

You use these symbols to first define $\bar{R}$ as follows:

$$\bar{R} = \frac{\sum\limits_{i=1}^{k} R_i}{k} = \frac{\text{sum of the ranges}}{\text{number of subgroups}}$$

You then use this symbol to write the equations for the control limits:

$$LCL = \bar{R} - 3\bar{R}\frac{d_3}{d_2} \qquad UCL = \bar{R} + 3\bar{R}\frac{d_3}{d_2}$$

in which the symbols d_3 and d_2 represent control chart factors obtained from Table C.5.

The equations can be simplified by using the D_3 **factor that is** equal to $1 - 3(d_3/d_2)$ and the D_4 **factor**, equal to $1 + 3(d_3/d_2)$. Values of D_3 and D_4 for different subgroup sizes are listed in Table C.5. The simplified equations are as follows:

$$LCL = D_3\bar{R} \qquad UCL = D_4\bar{R}$$

For the bank waiting times data, the sum of the ranges is 65.5 and the number of groups is 20. Therefore,

$$\bar{R} = \frac{\text{sum of the ranges}}{\text{number of groups}}$$

$$= \frac{65.5}{20} = 3.275$$

For a subgroup size = 4, $D_3 = 0$ and $D_4 = 2.282$

$$LCL = (0)(3.275) = 0$$

$$UCL = (2.282)(3.275) = 7.4736$$

Equations for the Lower and Upper Control Limits for the Mean

You use the following symbols to write the equations for the lower and upper control limits for the mean:

- A subscripted X Bar, $\bar{X}_i$, for the sample mean of n observations at time i,
- A subscripted X double bar, $\bar{\bar{X}}$, for the mean of all the sample means,
- A subscripted uppercase italic R, R_i, for the range of n observations at time i,
- A lowercase italic k, k, for the number of groups

You use these symbols to first define $\bar{\bar{X}}$ and $\bar{R}$ as

$$\bar{\bar{X}} = \frac{\sum_{i=1}^{k} \bar{X}_i}{k} = \frac{\text{sum of the sample means}}{\text{number of groups}}$$

and

$$\bar{R} = \frac{\sum_{i=1}^{k} R_i}{k} = \frac{\text{sum of the ranges}}{\text{number of groups}}$$

You then use these symbols to write the equations for the control limits:

$$LCL = \bar{\bar{X}} - 3\frac{\bar{R}}{d_2\sqrt{n}} \qquad UCL = \bar{\bar{X}} + 3\frac{\bar{R}}{d_2\sqrt{n}}$$

in which d_2 represents a control chart factor obtained from Table C.5. The equations can be simplified by using the A_2 factor that is equal to $3/(d_2\sqrt{n})$. Values of A_2 for different subgroup sizes are listed in Table C.5. The simplified equations are as follows:

$$LCL = \bar{\bar{X}} - A_2\bar{R} \qquad UCL = \bar{\bar{X}} + A_2\bar{R}$$

For the bank waiting time data, the sum of the ranges is 65.5, the sum of the sample means is 118.825, and the number of groups is 20. Therefore,

$$\overline{R} = \frac{\text{sum of the ranges}}{\text{number of groups}}$$

$$= \frac{65.5}{20} = 3.275$$

$$\overline{\overline{X}} = \frac{\text{sum of the sample means}}{\text{number of groups}}$$

$$= \frac{118.825}{20} = 5.94125$$

For a group size = 4, $A_2 = 0.729$

$LCL = 5.94125 - (0.729)\,(3.275) = 3.553775$

$UCL = 5.94125 + (0.729)\,(3.275) = 8.328725$

Important Equations

Lower and Upper Control Limits for the p chart

$$(12.1) \quad LCL = \overline{p} - 3\sqrt{\frac{\overline{p}(1-\overline{p})}{\overline{n}}}$$

$$(12.2) \quad UCL = \overline{p} + 3\sqrt{\frac{\overline{p}(1-\overline{p})}{\overline{n}}}$$

Lower and Upper Control Limits for the Range

$$(12.3) \quad LCL = \overline{R} - 3\overline{R}\frac{d_3}{d_2}$$

$$(12.4) \quad UCL = \overline{R} + 3\overline{R}\frac{d_3}{d_2}$$

$$(12.5) \quad LCL = D_3\overline{R}$$

$$(12.6) \quad UCL = D_4\overline{R}$$

Lower and Upper Control Limits for the Mean

$$(12.7) \quad LCL = \bar{\bar{X}} - 3 \frac{\bar{R}}{d_2 \sqrt{n}}$$

$$(12.8) \quad UCL = \bar{\bar{X}} + 3 \frac{\bar{R}}{d_2 \sqrt{n}}$$

$$(12.9) \quad LCL = \bar{\bar{X}} - A_2 \bar{R}$$

$$(12.10) \quad UCL = \bar{\bar{X}} + A_2 \bar{R}$$

One-Minute Summary

Quality management approaches

- Total quality management (TQM)
- Six Sigma DMAIC model

Process control techniques

- If a categorical variable, use attribute control charts such as p charts.
- If a continuous numerical variable, use variables control charts such as R and $\bar{X}$ charts.

Test Yourself
Short Answers

1. The control chart:
    (a) focuses on the time dimension of a system
    (b) captures the natural variability in the system
    (c) can be used for categorical or numerical variables
    (d) All of the above

2. Variation signaled by individual fluctuations or patterns in the data is called:
    (a) special causes of variation
    (b) common causes of variation

(c) Six Sigma

(d) the red bead experiment

3. Variation due to the inherent variability in a system of operation is called:
    (a) special causes of variation

    (b) common causes of variation

    (c) Six Sigma

    (d) the red bead experiment

4. Which of the following is not one of Deming's 14 points?
    (a) Believe in mass inspection

    (b) Create constancy of purpose for improvement of product or service

    (c) Adopt and institute leadership

    (d) Drive out fear

5. The principal focus of the control chart is the attempt to separate special or assignable causes of variation from common causes of variation. What cause of variation can be reduced only by changing the system?
    (a) Special or assignable causes

    (b) Common causes

    (c) Total causes

    (d) None of the above

6. After the control limits are set for a control chart, you attempt to:
    (a) discern patterns that might exist in values over time

    (b) determine whether any points fall outside the control limits

    (c) Both of the above

    (d) None of the above

7. Which of the following situations suggests a process that appears to be operating in a state of statistical control?
    (a) A control chart with a series of consecutive points that are above the center line and a series of consecutive points that are below the center line

    (b) A control chart in which no points fall outside either the upper control limit or the lower control limit and no patterns are present

    (c) A control chart in which several points fall outside the upper control limit

    (d) All of the above

8. Which of the following situations suggests a process that appears to be operating out of statistical control?

   (a) A control chart with a series of eight consecutive points that are above the center line

   (b) A control chart in which points fall outside the lower control limit

   (c) A control chart in which points fall outside the upper control limit

   (d) All of the above

9. A process is said to be out of control if:

   (a) a point falls above the upper control limits or below the lower control limits

   (b) eight or more consecutive points are above the center line

   (c) Either (a) or (b)

   (d) Neither (a) or (b)

10. One of the morals of the red bead experiment is:

    (a) variation is part of the process

    (b) only management can change the system

    (c) it is the system that primarily determines performance

    (d) All of the above

11. The cause of variation that can be reduced only by changing the system is _____ cause variation.

12. _____ causes of variation are correctable without modifying the system.

Answer True or False:

13. The control limits are based on the standard deviation of the process.

14. The purpose of a control chart is to eliminate common cause variation.

15. Special causes of variation are signaled by individual fluctuations or patterns in the data.

16. Common causes of variation represent variation due to the inherent variability in the system.

17. Common causes of variation are correctable without modifying the system.

18. Changes in the system to reduce common cause variation are the responsibility of management.

19. The *p* chart is a control chart used for monitoring the proportion of items that have a certain characteristic.

20. It is not possible for the $\overline{X}$ chart to be out of control when the R chart is in control.

21. One of the morals of the red bead experiment is that variation is part of any process.

22. The R chart is a control chart used to monitor a process mean.

23. Developing operational definitions for each critical-to-quality characteristic involves the Define part of the DMAIC process.

24. Studying the importance of each process variable on the Critical-To-Quality (CTQ) characteristic using designed experiments involves the Measure part of the DMAIC process.

Answers to Test Yourself Short Answers

1. d	13. True
2. a	14. False
3. b	15. True
4. a	16. True
5. b	17. False
6. c	18. True
7. b	19. True
8. d	20. False
9. c	21. True
10. d	22. False
11. common	23. True
12. special	24. False

Problems

1. A medical transcription service enters medical data on patient files for hospitals. The service studied ways to improve the turnaround time (defined as the time between sending data and the time the client receives completed files). After studying the process, the service determined that turnaround time was increased by transmission errors. A transmission error was defined as data transmitted that did not go through as planned and needed to be retransmitted. Each day, a sample of 125 transmissions was randomly selected and evaluated for errors. The following table presents the number and proportion of transmissions with errors:

Transmit

Day (i)	Number of Errors (X_i)	Proportion of Errors (p_i)	Day (i)	Number of Errors (X_i)	Proportion of Errors (p_i)
1	6	0.048	17	4	0.032
2	3	0.024	18	6	0.048
3	4	0.032	19	3	0.024
4	4	0.032	20	5	0.040
5	9	0.072	21	1	0.008
6	0	0.000	22	3	0.024
7	0	0.000	23	14	0.112
8	8	0.064	24	6	0.048
9	4	0.032	25	7	0.056
10	3	0.024	26	3	0.024
11	4	0.032	27	10	0.080
12	1	0.008	28	7	0.056
13	10	0.080	29	5	0.040
14	9	0.072	30	0	0.000
15	3	0.024	31	3	0.024
16	1	0.008			

(a) Construct a p chart.

(b) Is the process in a state of statistical control? Why?

2. The bottling division of a soft drink company maintains daily records of the occurrences of unacceptable cans flowing from the filling and sealing machine at a bottling facility. The following table lists the number of cans filled and the number of nonconforming cans for one month (based on a five-day workweek):

Cola

Day	Cans Filled	Unacceptable Cans	Day	Cans Filled	Unacceptable Cans
1	5,043	47	12	5,314	70
2	4,852	51	13	5,097	64
3	4,908	43	14	4,932	59
4	4,756	37	15	5,023	75
5	4,901	78	16	5,117	71

Day	Cans Filled	Unacceptable Cans	Day	Cans Filled	Unacceptable Cans
6	4,892	66	17	5,099	68
7	5,354	51	18	5,345	78
8	5,321	66	19	5,456	88
9	5,045	61	20	5,554	83
10	5,113	72	21	5,421	82
11	5,247	63	22	5,555	87

(a) Construct a p chart for the proportion of unacceptable cans for the month. Does the process give an out-of-control signal?

(b) If you want to develop a process for reducing the proportion of unacceptable cans, how should you proceed?

3. The director of radiology at a large metropolitan hospital is concerned about scheduling in the radiology facilities. On a typical day, 250 patients are transported to the radiology department for treatment or diagnostic procedures. If patients do not reach the radiology unit at their scheduled times, backups occur, and other patients experience delays. The time it takes to transport patients to the radiology unit is operationally defined as the time between when the transporter is assigned to the patient and when the patient arrives at the radiology unit. A sample of $n = 4$ patients was selected each day for 20 days, and the time to transport each patient (in minutes) was determined, with the results in the following table.

Transport

Day	Transport Times	Day	Transport Times	Day	Transport Times	Day	Transport Times
1	16.3	6	15.2	11	15.6	16	9.7
1	17.4	6	23.6	11	19.1	16	14.6
1	18.7	6	19.4	11	22.9	16	10.4
1	16.9	6	20.0	11	19.4	16	10.8
2	29.4	7	23.1	12	19.8	17	27.8
2	17.3	7	13.6	12	12.2	17	18.4
2	22.7	7	21.1	12	26.7	17	23.7
2	10.9	7	13.7	12	19.0	17	22.8
3	12.2	8	15.7	13	24.3	18	17.4

(continues)

Day	Transport Times	Day	Transport Times	Day	Transport Times	Day	Transport Times
3	12.7	8	10.9	13	18.7	18	25.8
3	14.1	8	16.4	13	30.3	18	18.4
3	10.3	8	21.8	13	22.9	18	9.0
4	22.4	9	10.2	14	16.5	19	20.5
4	19.7	9	14.9	14	14.3	19	17.8
4	24.9	9	12.6	14	19.5	19	23.2
4	23.4	9	11.9	14	15.5	19	18.0
5	13.5	10	14.7	15	23.4	20	14.2
5	11.6	10	18.7	15	27.6	20	14.6
5	14.8	10	22.0	15	30.7	20	11.1
5	13.5	10	19.1	15	24.0	20	17.7

(a) Construct control charts for the range and the mean.

(b) Is the process in control?

4. A print production team of a newspaper is charged with improving the quality of the newspaper. The team has chosen the blackness of the print of the newspaper as its first project. Blackness is measured on a device that records the results on a standard scale, where the blacker the spot, the larger the blackness measure. The blackness of the print should be approximately 1.0. Five spots on the first newspaper printed each day are randomly selected, and the blackness of each spot is measured. The following table presents the results for 25 days.

Newsprint Blackness for 25 Consecutive Days

Newspaper

Day	Spot Number 1	2	3	4	5
1	0.96	1.01	1.12	1.07	0.97
2	1.06	1.00	1.02	1.16	0.96
3	1.00	0.90	0.98	1.18	0.96
4	0.92	0.89	1.01	1.16	0.90
5	1.02	1.16	1.03	0.89	1.00
6	0.88	0.92	1.03	1.16	0.91
7	1.05	1.13	1.01	0.93	1.03

| | | Spot Number | | |
Day	1	2	3	4	5
8	0.95	0.86	1.14	0.90	0.95
9	0.99	0.89	1.00	1.15	0.92
10	0.89	1.18	1.03	0.96	1.04
11	0.97	1.13	0.95	0.86	1.06
12	1.00	0.87	1.02	0.98	1.13
13	0.96	0.79	1.17	0.97	0.95
14	1.03	0.89	1.03	1.12	1.03
15	0.96	1.12	0.95	0.88	0.99
16	1.01	0.87	0.99	1.04	1.16
17	0.98	0.85	0.99	1.04	1.16
18	1.03	0.82	1.21	0.98	1.08
19	1.02	0.84	1.15	0.94	1.08
20	0.90	1.02	1.10	1.04	1.08
21	0.96	1.05	1.01	0.93	1.01
22	0.89	1.04	0.97	0.99	0.95
23	0.96	1.00	0.97	1.04	0.95
24	1.01	0.98	1.04	1.01	0.92
25	1.01	1.00	0.92	0.90	1.11

(a) Construct control charts for the range and the mean.
(b) Is the process in a state of statistical control? Explain.

Answers to Test Yourself Problems

1. (a) $LCL = -0.0134 < 0$, so LCL does not exist. $UCL = 0.0888$.
   (b) The proportion of transmissions with errors on Day 23 is substantially out of control. Possible causes of this value should be investigated.

2. (a) $UCL = 0.0176$, $LCL = 0.0082$.
   (b) The proportion of unacceptable cans is below the LCL on Day 4. There is evidence of a pattern over time because the last eight points are all above the mean and most of the earlier points are below the mean. Therefore, this process is out of control. You

need to find the reasons for the special cause variation and take corrective action.

3. (a) $\bar{R}$ = 8.145, LCL does not exist, UCL = 18.5869; $\bar{\bar{X}}$ = 18.12, UCL = 24.0577, LCL = 12.1823.

(b) No sample ranges are outside the control limits and there does not appear to be a pattern in the range chart. The mean is above the UCL on Day 15 and below the LCL on Day 16. Therefore, the process is not in control.

4. (a) $\bar{R}$ = 0.2424, UCL = 0.5124, and LCL does not exist; $\bar{\bar{X}}$ = 1.00032, UCL = 1.14022, and LCL = 0.86042.

(b) There are no points outside the control limits on the range chart and no other violations of the rules. There are no points outside the control limits on the $\bar{X}$ chart and no other violations of the rules.

References

1. Arndt, M., "Quality Isn't Just for Widgets," *BusinessWeek*, July 22, 2002, 72–73.

2. Berenson, M. L., D. M. Levine, and T. C. Krehbiel. *Basic Business Statistics: Concepts and Applications*, Eleventh Edition. Upper Saddle River, NJ: Prentice Hall, 2009.

3. Cyger, M. "The Last Word—Riding the Bandwagon," *iSixSigma Magazine*, November/December 2006.

4. Deming, W. E., *The New Economics for Business, Industry, and Government* (Cambridge, MA: MIT Center for Advanced Engineering Study, 1993).

5. Deming, W. E., *Out of the Crisis* (Cambridge, MA: MIT Center for Advanced Engineering Study, 1986).

6. Gabor, A., *The Man Who Discovered Quality* (New York: Time Books, 1990).

7. Gitlow, H., and D. Levine, *Six Sigma for Green Belts and Champions* (Upper Saddle River, NJ: Financial Times/ Prentice Hall, 2005).

8. Gitlow, H., D. Levine, and E. Popovich, *Design for Six Sigma for Green Belts and Champions* (Upper Saddle River, NJ: Financial Times/Prentice Hall, 2006).

9. Halberstam, D. *The Reckoning*. New York: Morrow, 1986.

10. Levine, D. M., *Statistics for Six Sigma for Green Belts with Minitab and JMP* (Upper Saddle River, NJ: Financial Times/Prentice Hall, 2006).

11. Levine, D. M., D. Stephan, T. C. Krehbiel, and M. L. Berenson. *Statistics for Managers Using Microsoft Excel*, Fifth Edition. Upper Saddle River, NJ: Prentice Hall, 2008.

12. Levine, D. M., P. C. Ramsey, and R. K. Smidt. *Applied Statistics for Engineers and Scientists Using Microsoft Excel and Minitab*. Upper Saddle River, NJ: Prentice Hall, 2001.

13. Snee, R. D., "Impact of Six Sigma on Quality," *Quality Engineering*, 12 (2000): ix–xiv.

14. Walton, M. *The Deming Management Method.* New York: Perigee Books, Putnam Publishing Group, 1986.

Calculator and Spreadsheet Operation and Configuration

Use this appendix to learn how to operate and configure your spreadsheet or calculator for use with this book. This appendix also reviews the conventions used in the book to describe various calculator and spreadsheet user operations.

Calculator Operation Conventions

Keystroke Conventions

The Calculator Keys sections of this book present keystroke-by-keystroke instructions for using your TI-83 or T1-84 family calculator. Individual keys are named by their primary legend and enclosed in a pair of brackets. For example, [2nd] refers to the colored key that starts the second row of keys on your calculator, whereas [ENTER] refers to the key at or near the lower-right corner of your calculator.

To refer to the four cursor keys that are immediately above the [VARS] and [CLEAR] keys, this book uses [◀], [▶], [▲], and [▼]. An instruction that reads "press [STAT][◀]" means press the STAT key followed by the left cursor key.

Note: Some Texas Instruments instructional materials name the secondary legends of keys in their instructions. For example, whereas instructions in this book would always state "press [2nd][STAT]" to display the List menu, some TI materials would say "press [2nd][LIST]" (LIST is the secondary legend of the STAT key).

Calculator Initial State

All instructions assume that you are beginning from the home screen and are not in the middle of some calculator activity. Usually, pressing [CLEAR] will clear your screen and place you on the home screen with a blinking cursor. (For some procedures, you will need to press [2nd][MODE] to return to the home screen.)

Using Menus

The instructions in this book always use menus to select statistical operations. If you are an advanced user who prefers typing command lines to choose a statistical function, you can use that method instead.

Selecting Menus

When you press a key such as the STAT key, you are presented with two or more menu screens. In this book, many instruction sequences contain phrases such as "press [STAT][◀] to display the Tests menu" that are shortcut ways of selecting a necessary menu screen. For example, pressing [STAT] can lead to the (Stat) Edit, Calc, or Tests menus. When you see a sequence such as "press [STAT][◀]," make sure you press the cursor key in order to display the correct screen. If you make a mistake, you can press [CLEAR] to start over.

Menu List Choices

When the instructions require you to make choices from an onscreen menu list, this book always states the instruction in the form "select *n:Choice* and press [ENTER]." You use the down cursor key ([▼]) to highlight the choice and then press the [ENTER] key to execute this instruction. If you prefer, you can press the key that corresponds to the *n* value *without* pressing [ENTER]. For example, given the instruction "select **6:2-PropZTest** and press [ENTER]" you could press [6] to immediately jump to the 2-PropZTest screen. As you gain experience selecting tests, you might prefer this method of making a menu list choice.

Busy Screen

Some instructions mention that you will see a "busy screen." A busy screen is one in which the calculator's **busy indicator**, a running vertical line in the top-right corner of the screen, is being displayed.

ac2 Calculator Technical Configuration

Memory Initial State

For most Calculator Keys sections, the initial memory state of your calculator is not critical. However, sometimes you may need to **reset** (clear) the memory of the calculator or reset the calculator to its factory default settings. To perform a calculator reset, press [2nd][+] to display the MEMORY screen. Select 7:**Reset** and press [ENTER] to display the RAM Reset screen. From this screen, press [1][2] to reset RAM memory or press [2][2] to reset the calculator to its factory defaults.

Numeric Notation Settings

Instructions in this book were developed with a calculator set to **Normal** numeric notation and **floating decimal** numeric format. To set your calculator to these settings (or to verify them):

1. Press [MODE] and then select **Normal** and press [ENTER].
2. Press [▼] and then select **Float** and press [ENTER].
3. Press [2nd][MODE] to return to the main screen.

Diagnostics Settings

You will need to turn on "calculator diagnostics" to ensure the proper functioning of certain advanced procedures such as regression. To turn on (or verify that diagnostics are turned on), press [2nd][0] to display the CATA-LOG screen and then select **DiagnosticOn** and press [ENTER]. When the DiagnosticOn prompt appears onscreen, press [ENTER] a second time to turn the diagnostics on.

Variable Data Value or Program Deletion

To free up memory or ensure the proper working of the accessory program, you can delete the data values that have been stored in a calculator variable that you no longer need. To do this, press [2nd][+] to display the MEMORY screen and then select 2:**Mem Mgmt/Del** and press [ENTER]. On the next screen, select a choice; for example, 1:**All** and press [ENTER]. Scroll through the list using the cursor keys to highlight a variable name and then press [DEL] to delete that variable's data values. Although 1:**All** displays the complete list of variances, 4:**List**, 5:**Matrix**, 6:**Y-Vars**, and 7:**Prgm** can be more useful choices when using this book.

ac3 **Using the A2MULREG Program**

The Calculator Keys section in Chapter 11, "Multiple Regression," uses the
A2MULREG program to perform multiple regression analysis. (This program
can also be used for simple linear regression analysis discussed in Chapter 10,
"Simple Linear Regression.") To use this program, you must download it from
the Texas Instruments Education Technology website (**education.ti.com**). Use
the website search function to search **A2MULREG** to find the web page that
contains the download link as well as full instructions for downloading the
program to your computer and then transferring it to your calculator (using
one of the Texas Instruments linking cables that is packaged with your calculator or available separately).

When you follow the instructions and successfully transfer the program to
your calculator, you can place the program in the user data archive (ARC). If
your copy is so archived, you need to unarchive it before you can use it for
regression analysis. To unarchive the program, press [**2nd**][**+**] to display the
MEMORY screen and then select **6:UnArchive** and press [**ENTER**]. At the
UnArchive prompt, press [**PGRM**] and then select the **A2MULREG** choice
and press [**ENTER**]. On the MEMORY screen, a 5:Archive choice can prove
useful if you want to later save the **A2MULREG** program before clearing
RAM memory.

If you have previously downloaded and transferred programs (sometimes
called Apps in Texas Instruments materials), you can delete one or more to
make room for **A2MULREG**. To delete a program, follow the instructions in
the preceding "Variable Data Value or Program Deletion" section and select
7:Prgm from the MEMORY screen to view the programs you can delete.

ac4 **Using TI Connect**

To transfer the **A2MULREG** program used in Chapter 11, you need to install
the **TI Connect** program that is available on the CD-ROM packaged with
your calculator or that you can download from the Texas Instruments
Education Technology website (**education.ti.com**).

This program has many other functions that you can find useful while using
this book. For example, the **TI DataEditor** component gives you an alternative way of entering data values for matrix variables (used in Chapter 9,
"Hypothesis Testing: Chi-Square Tests and the One-Way Analysis of Variance
(ANOVA)" and Chapter 11) as well as list variables (used throughout this
book). In particular, the editor provides you with a handy way of transferring
worksheet data values and assigning them to list or matrix variables (through

a simple copy-and-paste operation). You can also use **Backup and Restore** to back up and restore the contents of your calculator, including all preloaded programs (Apps) and **TI DeviceExplorer** or **TI DeviceInfo** to learn more about the status of your calculator. If you want to "save" your results screen for later use, consider using **TI ScreenCapture**, the method used to capture and display calculator screens in this book.

Full instructions for using these components or **TI Connect** are available in the TI Connect help system.

A.S1 Spreadsheet Operation Conventions

The spreadsheet operation instructions in this book use a standard vocabulary to describe keystroke and mouse (pointer) operations. Keys are always named by their legends. For example, the instruction "press Enter" means to press the key with the legend **Enter**.

For mouse operation, this book uses **click** and **select** and less frequently, **check**, **right-click**, and **double-click**. Click means to move the pointer over an object and press the primary mouse button. Select means to either find and highlight a named choice from a pull-down list or fill in an option (also known as radio) button associated with that choice. Check means to fill in the check box of a choice by clicking in its empty check box. Right-click means to press the secondary mouse button (or to hold down the Control key and press the mouse button, if using a one-button mouse). Double-click means to press the primary mouse button rapidly twice to select an object directly.

A.S2 Spreadsheet Technical Configurations

The instructions in this book for using Microsoft Excel and OpenOffice.org Calc 3 assume no special technical settings. If you plan to use any of the Analysis ToolPak Tips of Appendix E, "Advanced Techniques," (Microsoft Excel only), you need to make sure that the Analysis ToolPak add-in has been installed in your copy of Microsoft Excel. (The Analysis ToolPak is *not included* with and is *not available* for Mac Excel 2008.)

To check whether the Analysis ToolPak is installed in your copy of Excel 97–2003:

1. Open Microsoft Excel.
2. Select **Tools** and then **Add-Ins**.

3. In the Add-Ins dialog box, verify that the **Analysis ToolPak** appears in the **Add-Ins available** list. Check the item if it is not already checked and then click **OK**.

4. Exit Microsoft Excel (to save the selections).

To check whether the Analysis ToolPak is installed in your copy of Excel 2007 or a later Windows Excel version:

1. Open Microsoft Excel.

2. Click the **Office Button**.

3. In the Office Button pane, click **Excel Options**.

4. In the Excel Options dialog box, click **Add-Ins** in the left pane and then look for **Analysis ToolPak** in the **Active Application Add-ins** list in the right pane. (Should **Analysis ToolPak** appear instead in the **Inactive Application Add-ins**, click **Go**. In the Add-Ins dialog box, check **Analysis ToolPak** and then click **OK**.)

5. Click **OK** (in the Excel Options dialog box) to close the window.

If **Analysis ToolPak** does not appear in your **Add-Ins available** list, you need to rerun the Microsoft Excel (or Office) setup program using your original Microsoft Office/Excel CD-ROM or DVD to add this component to your Excel copy.

APPENDIX

B

Review of Arithmetic and Algebra

The authors understand and realize that wide differences exist in the mathematical background of readers of this book. Some of you might have taken various courses in algebra, calculus, and matrix algebra, while others might not have taken any mathematics courses in a long period of time. Because the emphasis in this book is on statistical concepts and the interpretation of spreadsheet and statistical calculator results, no prerequisite beyond elementary algebra is needed. To assess your arithmetic and algebraic skills, answer the following questions and then read the review that follows.

Assessment Quiz

Part 1

Fill in the correct answer.

1. $\dfrac{\frac{1}{2}}{3} =$

2. $(0.4)^2 =$

3. $1 + \dfrac{2}{3} =$

4. $\left(\dfrac{1}{3}\right)^{(4)} =$

5. $\dfrac{1}{5} =$ (in decimals)

6. $1 - (-0.3) =$

7. $4 \times 0.2 \times (-8) =$

8. $\left(\dfrac{1}{4} \times \dfrac{2}{3} \right) =$

9. $\left(\dfrac{1}{100} \right) + \left(\dfrac{1}{200} \right) =$

10. $\sqrt{16} =$

Part 2

Select the correct answer.

1. If $a = bc$, then $c =$
   a. ab
   b. b/a
   c. a/b
   d. none of the above

2. If $x + y = z$, then $y =$
   a. z/x
   b. $z + x$
   c. $z - x$
   d. none of the above

3. $(x^3)(x^2) =$
   a. x^5
   b. x^6
   c. x^1
   d. none of the above

4. $x^0 =$
   a. x
   b. 1
   c. 0
   d. none of the above

5. $x(y - z) =$
    a. $xy - xz$
    b. $xy - z$
    c. $(y - z)/x$
    d. none of the above

6. $(x + y)/z =$
    a. $(x/z) + y$
    b. $(x/z) + (y/z)$
    c. $x + (y/z)$
    d. none of the above

7. $x /(y + z) =$
    a. $(x/y) + (1/z)$
    b. $(x/y) + (x/z)$
    c. $(y + z)/ x$
    d. none of the above

8. If $x = 10$, $y = 5$, $z = 2$, and $w = 20$, then $(xy - z^2)/w =$
    a. 5
    b. 2.3
    c. 46
    d. none of the above

9. $(8x^4)/(4x^2) =$
    a. $2x^2$
    b. 2
    c. $2x$
    d. none of the above

10. $\sqrt{\dfrac{X}{Y}} =$

    a. $\sqrt{Y}/\sqrt{X}$

    b. $\sqrt{1}/\sqrt{XY}$

    c. $\sqrt{X}/\sqrt{Y}$

    d. none of the above

The answers to both parts of the quiz appear at the end of this appendix.

Symbols

Each of the four basic arithmetic operations—addition, subtraction, multiplication, and division—is indicated by a symbol.

+ add

× or • multiply

– subtract

÷ or / divide

In addition to these operations, the following symbols are used to indicate equality or inequality

= equals

≠ not equal

≅ approximately equal to

> greater than

< less than

≥ greater than or equal to

≤ less than or equal to

Addition

Addition refers to the summation or accumulation of a set of numbers. In adding numbers, the two basic laws are the commutative law and the associative law.

The **commutative law** of addition states that the order in which numbers are added is irrelevant. This can be seen in the following two examples.

$$1 + 2 = 3 \qquad 2 + 1 = 3$$
$$x + y = z \qquad y + x = z$$

In each example, the number that was listed first and the number that was listed second did not matter.

The **associative law** of addition states that in adding several numbers, any subgrouping of the numbers can be added first, last, or in the middle. You can see this in the following examples:

$$2 + 3 + 6 + 7 + 4 + 1 = 23$$
$$(5) + (6 + 7) + 4 + 1 = 23$$
$$5 + 13 + 5 = 23$$
$$5 + 6 + 7 + 4 + 1 = 23$$

In each of these examples, the order in which the numbers have been added has no effect on the results.

Subtraction

The process of subtraction is the opposite or inverse of addition. The operation of subtracting 1 from 2 (that is, $2 - 1$) means that one unit is to be taken away from two units, leaving a remainder of one unit. In contrast to addition, the commutative and associative laws do not hold for subtraction. Therefore, as indicated in the following examples,

$8 - 4 = 4$	but	$4 - 8 = -4$
$3 - 6 = -3$	but	$6 - 3 = 3$
$8 - 3 - 2 = 3$	but	$3 - 2 - 8 = -7$
$9 - 4 - 2 = 3$	but	$2 - 4 - 9 = -11$

When subtracting negative numbers, remember that the same result occurs when subtracting a negative number as when adding a positive number. Thus,

$4 - (-3) = +7$	$4 + 3 = 7$
$8 - (-10) = +18$	$8 + 10 = 18$

Multiplication

The operation of multiplication is a shortcut method of addition when the same number is to be added several times. For example, if 7 is added three times ($7 + 7 + 7$), you could multiply 7 times 3 to get the product of 21.

In multiplication as in addition, the commutative laws and associative laws are in operation so that:

$$a \times b = b \times a$$
$$4 \times 5 = 5 \times 4 = 20$$
$$(2 \times 5) \times 6 = 10 \times 6 = 60$$

A third law of multiplication, the **distributive law**, applies to the multiplication of one number by the sum of several numbers. Here,

$$a(b + c) = ab + ac$$
$$2(3 + 4) = 2(7) = 2(3) + 2(4) = 14$$

The resulting product is the same regardless of whether b and c are summed and multiplied by a, or a is multiplied by b and by c and the two products are added together.

You also need to remember that when multiplying negative numbers, a negative number multiplied by a negative number equals a positive number. Thus,

$$(-a) \times (-b) = ab$$
$$(-5) \times (-4) = +20$$

Division

Just as subtraction is the opposite of addition, division is the opposite or inverse of multiplication. Division can be viewed as a shortcut to subtraction. When you divide 20 by 4, you are actually determining the number of times that 4 can be subtracted from 20. In general, however, the number of times one number can be divided by another may not be an exact integer value because there could be a remainder. For example, if you divide 21 by 4, the answer is 5 with a remainder of 1, or $5\frac{1}{4}$.

As in the case of subtraction, neither the commutative nor associative law of addition and multiplication holds for division.

$$a \div b \neq b \div a$$
$$9 \div 3 \neq 3 \div 9$$
$$6 \div (3 \div 2) = 4$$
$$(6 \div 3) \div 2 = 1$$

The distributive law holds only when the numbers to be added are contained in the numerator, not the denominator. Thus,

$$\frac{a+b}{c} = \frac{a}{c} + \frac{b}{c} \quad \text{but} \quad \frac{a}{b+c} \neq \frac{a}{b} + \frac{a}{c}$$

For example,

$$\frac{1}{2+3} = \frac{1}{5} \quad \text{but} \quad \frac{1}{2+3} \neq \frac{1}{2} + \frac{1}{3}$$

The last important property of division states that if the numerator and the denominator are multiplied or divided by the same number, the resulting quotient is not affected. Therefore,

$$\frac{80}{40} = 2$$

then

$$\frac{5(80)}{5(40)} = \frac{400}{200} = 2$$

and

$$\frac{80 \div 5}{40 \div 5} = \frac{16}{8} = 2$$

Fractions

A fraction is a number that consists of a combination of whole numbers and/or parts of whole numbers. For instance, the fraction $1/3$ consists of only one portion of a number, while the fraction $7/6$ consists of the whole number 1 plus the fraction $1/6$. Each of the operations of addition, subtraction, multiplication, and division can be used with fractions. When adding and subtracting fractions, you must find the lowest common denominator for each fraction prior to adding or subtracting them. Thus, in adding $1/3 + 1/5$, the lowest common denominator is 15, so

$$\frac{5}{15} + \frac{3}{15} = \frac{8}{15}$$

In subtracting $1/4 - 1/6$, the same principles applies, so that the lowest common denominator is 12, producing a result of

$$\frac{3}{12} - \frac{2}{12} = \frac{1}{12}$$

Multiplying and dividing fractions does not have the lowest common denominator requirement associated with adding and subtracting fractions. Thus, if a/b is multiplied by c/d, the result is ac/bd.

The resulting numerator, *ac*, is the product of the numerators *a* and *c*, while the denominator, *bd*, is the product of the two denominators *b* and *d*. The resulting fraction can sometimes be reduced to a lower term by dividing the numerator and denominator by a common factor. For example, taking

$$\frac{2}{3} \times \frac{6}{7} = \frac{12}{21}$$

and dividing the numerator and denominator by 3 produces the result $4/7$.

Division of fractions can be thought of as the inverse of multiplication, so the divisor can be inverted and multiplied by the original fraction. Thus,

$$\frac{9}{5} \div \frac{1}{4} = \frac{9}{5} \times \frac{4}{1} = \frac{36}{5}$$

The division of a fraction can also be thought of as a way of converting the fraction to a decimal number. For example, the fraction $2/5$ can be converted to a decimal number by dividing its numerator, 2, by its denominator, 5, to produce the decimal number 0.40.

Exponents and Square Roots

Exponentiation (raising a number to a power) provides a shortcut in writing numerous multiplications. For example, $2 \times 2 \times 2 \times 2 \times 2$ can be written as $2^5 = 32$. The 5 represents the exponent (or power) of the number 2, telling you that 2 is to be multiplied by itself five times.

Several rules can be used for multiplying or dividing numbers that contain exponents.

Rule 1 $x^a \cdot x^b = x^{(a + b)}$

If two numbers involving a power of the same number are multiplied, the product is the same number raised to the sum of the powers.

$$4^2 \cdot 4^3 = (4 \cdot 4)(4 \cdot 4 \cdot 4) = 4^5$$

Rule 2 $(x^a)^b = x^{ab}$

If you take the power of a number that is already taken to a power, the result is a number that is raised to the product of the two powers. For example,

$$(4^2)^3 = (4^2)(4^2)(4^2) = 4^6$$

Rule 3 $\dfrac{x^a}{x^b} = x^{(a-b)}$

If a number raised to a power is divided by the same number raised to a power, the quotient is the number raised to the difference of the powers. Thus

$$\frac{3^5}{3^3} = \frac{3 \cdot 3 \cdot 3 \cdot 3 \cdot 3}{3 \cdot 3 \cdot 3} = 3^2$$

If the denominator has a higher power than the numerator, the resulting quotient is a negative power. Thus,

$$\frac{3^3}{3^5} = \frac{3 \cdot 3 \cdot 3}{3 \cdot 3 \cdot 3 \cdot 3 \cdot 3} = \frac{1}{3^2} = 3^{-2} = \frac{1}{9}$$

If the difference between the powers of the numerator and denominator is 1, the result is the number itself. In other words, $x^1 = x$. For example,

$$\frac{3^3}{3^2} = \frac{3 \cdot 3 \cdot 3}{3 \cdot 3} = 3^1 = 3$$

If, however, no difference exists in the power of the numbers in the numerator and denominator, the result is 1. Thus,

$$\frac{x^a}{x^a} = x^{a-a} = x^0 = 1$$

Therefore, any number raised to the zero power equals 1. For example,

$$\frac{3^3}{3^3} = \frac{3 \cdot 3 \cdot 3}{3 \cdot 3 \cdot 3} = 3^0 = 1$$

The square root represented by the symbol $\sqrt{}$ is a special power of a number, the $\frac{1}{2}$ power. It indicates the value that when multiplied by itself, will produce the original number.

Equations

In statistics, many formulas are expressed as equations where one unknown value is a function of another value. Thus, it is important that you know how to manipulate equations into various forms. The rules of addition, subtraction, multiplication, and division can be used to work with equations. For example, the equation

$$x - 2 = 5$$

can be solved for x by adding 2 to each side of the equation. This results in $x - 2 + 2 = 5 + 2$. Therefore $x = 7$.

If

$$x + y = z$$

you could solve for x by subtracting y from both sides of the equation so that

$x + y - y = z - y$ Therefore $x = z - y$.

If the product of two variables is equal to a third variable, such as

$$x\,y = z,$$

you can solve for x by dividing both sides of the equation by y. Thus

$$\frac{xy}{y} = \frac{z}{y}$$

$$x = \frac{z}{y}$$

Conversely, if

$$\frac{x}{y} = z$$

you can solve for x by multiplying both sides of the equation by y.

$$\frac{xy}{y} = zy$$

$$x = zy$$

To summarize, the various operations of addition, subtraction, multiplication, and division can be applied to equations as long as the same operation is performed on each side of the equation, thereby maintaining the equality.

Answers to Quiz
Part 1

1. $\frac{3}{2}$
2. 0.16
3. $\frac{5}{3}$
4. $\frac{1}{81}$
5. 0.20
6. 1.30
7. −6.4
8. $+\frac{1}{6}$
9. $\frac{3}{200}$
10. 4

Part 2

1. c
2. c
3. a
4. b
5. a
6. b
7. d
8. b
9. a
10. c

APPENDIX

C

Statistical Tables

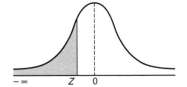

TABLE C.1

The Cumulative Standardized Normal Distribution

Entry represents area under the cumulative standardized normal distribution from −∞ to Z.

Z	0.00	0.01	0.02	0.03	0.04	0.05	0.06	0.07	0.08	0.09
-3.9	0.00005	0.00005	0.00004	0.00004	0.00004	0.00004	0.00004	0.00004	0.00003	0.00003
-3.8	0.00007	0.00007	0.00007	0.00006	0.00006	0.00006	0.00006	0.00005	0.00005	0.00005
-3.7	0.00011	0.00010	0.00010	0.00010	0.00009	0.00009	0.00008	0.00008	0.00008	0.00008
-3.6	0.00016	0.00015	0.00015	0.00014	0.00014	0.00013	0.00013	0.00012	0.00012	0.00011
-3.5	0.00023	0.00022	0.00022	0.00021	0.00020	0.00019	0.00019	0.00018	0.00017	0.00017
-3.4	0.00034	0.00032	0.00031	0.00030	0.00029	0.00028	0.00027	0.00026	0.00025	0.00024
-3.3	0.00048	0.00047	0.00045	0.00043	0.00042	0.00040	0.00039	0.00038	0.00036	0.00035
-3.2	0.00069	0.00066	0.00064	0.00062	0.00060	0.00058	0.00056	0.00054	0.00052	0.00050
-3.1	0.00097	0.00094	0.00090	0.00087	0.00084	0.00082	0.00079	0.00076	0.00074	0.00071
-3.0	0.00135	0.00131	0.00126	0.00122	0.00118	0.00114	0.00111	0.00107	0.00103	0.00100
-2.9	0.0019	0.0018	0.0018	0.0017	0.0016	0.0016	0.0015	0.0015	0.0014	0.0014
-2.8	0.0026	0.0025	0.0024	0.0023	0.0023	0.0022	0.0021	0.0021	0.0020	0.0019
-2.7	0.0035	0.0034	0.0033	0.0032	0.0031	0.0030	0.0029	0.0028	0.0027	0.0026
-2.6	0.0047	0.0045	0.0044	0.0043	0.0041	0.0040	0.0039	0.0038	0.0037	0.0036
-2.5	0.0062	0.0060	0.0059	0.0057	0.0055	0.0054	0.0052	0.0051	0.0049	0.0048
-2.4	0.0082	0.0080	0.0078	0.0075	0.0073	0.0071	0.0069	0.0068	0.0066	0.0064
-2.3	0.0107	0.0104	0.0102	0.0099	0.0096	0.0094	0.0091	0.0089	0.0087	0.0084
-2.2	0.0139	0.0136	0.0132	0.0129	0.0125	0.0122	0.0119	0.0116	0.0113	0.0110
-2.1	0.0179	0.0174	0.0170	0.0166	0.0162	0.0158	0.0154	0.0150	0.0146	0.0143
-2.0	0.0228	0.0222	0.0217	0.0212	0.0207	0.0202	0.0197	0.0192	0.0188	0.0183

TABLE C.1   313

Z	0.00	0.01	0.02	0.03	0.04	0.05	0.06	0.07	0.08	0.09
-1.9	0.0287	0.0281	0.0274	0.0268	0.0262	0.0256	0.0250	0.0244	0.0239	0.0233
-1.8	0.0359	0.0351	0.0344	0.0336	0.0329	0.0322	0.0314	0.0307	0.0301	0.0294
-1.7	0.0446	0.0436	0.0427	0.0418	0.0409	0.0401	0.0392	0.0384	0.0375	0.0367
-1.6	0.0548	0.0537	0.0526	0.0516	0.0505	0.0495	0.0485	0.0475	0.0465	0.0455
-1.5	0.0668	0.0655	0.0643	0.0630	0.0618	0.0606	0.0594	0.0582	0.0571	0.0559
-1.4	0.0808	0.0793	0.0778	0.0764	0.0749	0.0735	0.0721	0.0708	0.0694	0.0681
-1.3	0.0968	0.0951	0.0934	0.0918	0.0901	0.0885	0.0869	0.0853	0.0838	0.0823
-1.2	0.1151	0.1131	0.1112	0.1093	0.1075	0.1056	0.1038	0.1020	0.1003	0.0985
-1.1	0.1357	0.1335	0.1314	0.1292	0.1271	0.1251	0.1230	0.1210	0.1190	0.1170
-1.0	0.1587	0.1562	0.1539	0.1515	0.1492	0.1469	0.1446	0.1423	0.1401	0.1379
-0.9	0.1841	0.1814	0.1788	0.1762	0.1736	0.1711	0.1685	0.1660	0.1635	0.1611
-0.8	0.2119	0.2090	0.2061	0.2033	0.2005	0.1977	0.1949	0.1922	0.1894	0.1867
-0.7	0.2420	0.2388	0.2358	0.2327	0.2296	0.2266	0.2236	0.2206	0.2177	0.2148
-0.6	0.2743	0.2709	0.2676	0.2643	0.2611	0.2578	0.2546	0.2514	0.2482	0.2451
-0.5	0.3085	0.3050	0.3015	0.2981	0.2946	0.2912	0.2877	0.2843	0.2810	0.2776
-0.4	0.3446	0.3409	0.3372	0.3336	0.3300	0.3264	0.3228	0.3192	0.3156	0.3121
-0.3	0.3821	0.3783	0.3745	0.3707	0.3669	0.3632	0.3594	0.3557	0.3520	0.3483
-0.2	0.4207	0.4168	0.4129	0.4090	0.4052	0.4013	0.3974	0.3936	0.3897	0.3859
-0.1	0.4602	0.4562	0.4522	0.4483	0.4443	0.4404	0.4364	0.4325	0.4286	0.4247
-0.0	0.5000	0.4960	0.4920	0.4880	0.4840	0.4801	0.4761	0.4721	0.4681	0.4641

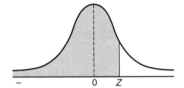

TABLE C.1 Continued

Entry represents area under the standardized normal distribution from $-\infty$ to Z.

Z	0.00	0.01	0.02	0.03	0.04	0.05	0.06	0.07	0.08	0.09
0.0	0.5000	0.5040	0.5080	0.5120	0.5160	0.5199	0.5239	0.5279	0.5319	0.5359
0.1	0.5398	0.5438	0.5478	0.5517	0.5557	0.5596	0.5636	0.5675	0.5714	0.5753
0.2	0.5793	0.5832	0.5871	0.5910	0.5948	0.5987	0.6026	0.6064	0.6103	0.6141
0.3	0.6179	0.6217	0.6255	0.6293	0.6331	0.6368	0.6406	0.6443	0.6480	0.6517
0.4	0.6554	0.6591	0.6628	0.6664	0.6700	0.6736	0.6772	0.6808	0.6844	0.6879
0.5	0.6915	0.6950	0.6985	0.7019	0.7054	0.7088	0.7123	0.7157	0.7190	0.7224
0.6	0.7257	0.7291	0.7324	0.7357	0.7389	0.7422	0.7454	0.7486	0.7518	0.7549
0.7	0.7580	0.7612	0.7642	0.7673	0.7704	0.7734	0.7764	0.7794	0.7823	0.7852
0.8	0.7881	0.7910	0.7939	0.7967	0.7995	0.8023	0.8051	0.8078	0.8106	0.8133
0.9	0.8159	0.8186	0.8212	0.8238	0.8264	0.8289	0.8315	0.8340	0.8365	0.8389
1.0	0.8413	0.8438	0.8461	0.8485	0.8508	0.8531	0.8554	0.8577	0.8599	0.8621
1.1	0.8643	0.8665	0.8686	0.8708	0.8729	0.8749	0.8770	0.8790	0.8810	0.8830
1.2	0.8849	0.8869	0.8888	0.8907	0.8925	0.8944	0.8962	0.8980	0.8997	0.9015
1.3	0.9032	0.9049	0.9066	0.9082	0.9099	0.9115	0.9131	0.9147	0.9162	0.9177
1.4	0.9192	0.9207	0.9222	0.9236	0.9251	0.9265	0.9279	0.9292	0.9306	0.9319
1.5	0.9332	0.9345	0.9357	0.9370	0.9382	0.9394	0.9406	0.9418	0.9429	0.9441
1.6	0.9452	0.9463	0.9474	0.9484	0.9495	0.9505	0.9515	0.9525	0.9535	0.9545
1.7	0.9554	0.9564	0.9573	0.9582	0.9591	0.9599	0.9608	0.9616	0.9625	0.9633
1.8	0.9641	0.9649	0.9656	0.9664	0.9671	0.9678	0.9686	0.9693	0.9699	0.9706
1.9	0.9713	0.9719	0.9726	0.9732	0.9738	0.9744	0.9750	0.9756	0.9761	0.9767
2.0	0.9772	0.9778	0.9783	0.9788	0.9793	0.9798	0.9803	0.9808	0.9812	0.9817
2.1	0.9821	0.9826	0.9830	0.9834	0.9838	0.9842	0.9846	0.9850	0.9854	0.9857
2.2	0.9861	0.9864	0.9868	0.9871	0.9875	0.9878	0.9881	0.9884	0.9887	0.9890

TABLE C.1   315

Z	0.00	0.01	0.02	0.03	0.04	0.05	0.06	0.07	0.08	0.09
2.3	0.9893	0.9896	0.9898	0.9901	0.9904	0.9906	0.9909	0.9911	0.9913	0.9916
2.4	0.9918	0.9920	0.9922	0.9925	0.9927	0.9929	0.9931	0.9932	0.9934	0.9936
2.5	0.9938	0.9940	0.9941	0.9943	0.9945	0.9946	0.9948	0.9949	0.9951	0.9952
2.6	0.9953	0.9955	0.9956	0.9957	0.9959	0.9960	0.9961	0.9962	0.9963	0.9964
2.7	0.9965	0.9966	0.9967	0.9968	0.9969	0.9970	0.9971	0.9972	0.9973	0.9974
2.8	0.9974	0.9975	0.9976	0.9977	0.9977	0.9978	0.9979	0.9979	0.9980	0.9981
2.9	0.9981	0.9982	0.9982	0.9983	0.9984	0.9984	0.9985	0.9985	0.9986	0.9986
3.0	0.99865	0.99869	0.99874	0.99878	0.99882	0.99886	0.99889	0.99893	0.99897	0.99900
3.1	0.99903	0.99906	0.99910	0.99913	0.99916	0.99918	0.99921	0.99924	0.99926	0.99929
3.2	0.99931	0.99934	0.99936	0.99938	0.99940	0.99942	0.99944	0.99946	0.99948	0.99950
3.3	0.99952	0.99953	0.99955	0.99957	0.99958	0.99960	0.99961	0.99962	0.99964	0.99965
3.4	0.99966	0.99968	0.99969	0.99970	0.99971	0.99972	0.99973	0.99974	0.99975	0.99976
3.5	0.99977	0.99978	0.99978	0.99979	0.99980	0.99981	0.99981	0.99982	0.99983	0.99983
3.6	0.99984	0.99985	0.99985	0.99986	0.99986	0.99987	0.99987	0.99988	0.99988	0.99989
3.7	0.99989	0.99990	0.99990	0.99990	0.99991	0.99991	0.99992	0.99992	0.99992	0.99992
3.8	0.99993	0.99993	0.99993	0.99994	0.99994	0.99994	0.99994	0.99995	0.99995	0.99995
3.9	0.99995	0.99995	0.99996	0.99996	0.99996	0.99996	0.99996	0.99996	0.99997	0.99997
4.0	0.99996832									
4.5	0.99999660									
5.0	0.99999971									
5.5	0.99999998									
6.0	0.99999999									

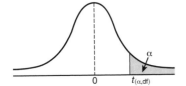

TABLE C.2

Critical Values of t

Degrees of Freedom	Upper-Tail Areas					
	0.25	0.10	0.05	0.025	0.01	0.005
1	1.0000	3.0777	6.3138	12.7062	31.8207	63.6574
2	0.8165	1.8856	2.9200	4.3027	6.9646	9.9248
3	0.7649	1.6377	2.3534	3.1824	4.5407	5.8409
4	0.7407	1.5332	2.1318	2.7764	3.7469	4.6041
5	0.7267	1.4759	2.0150	2.5706	3.3649	4.0322
6	0.7176	1.4398	1.9432	2.4469	3.1427	3.7074
7	0.7111	1.4149	1.8946	2.3646	2.9980	3.4995
8	0.7064	1.3968	1.8595	2.3060	2.8965	3.3554
9	0.7027	1.3830	1.8331	2.2622	2.8214	3.2498
10	0.6998	1.3722	1.8125	2.2281	2.7638	3.1693
11	0.6974	1.3634	1.7959	2.2010	2.7181	3.1058
12	0.6955	1.3562	1.7823	2.1788	2.6810	3.0545
13	0.6938	1.3502	1.7709	2.1604	2.6503	3.0123
14	0.6924	1.3450	1.7613	2.1448	2.6245	2.9768
15	0.6912	1.3406	1.7531	2.1315	2.6025	2.9467
16	0.6901	1.3368	1.7459	2.1199	2.5835	2.9208
17	0.6892	1.3334	1.7396	2.1098	2.5669	2.8982
18	0.6884	1.3304	1.7341	2.1009	2.5524	2.8784
19	0.6876	1.3277	1.7291	2.0930	2.5395	2.8609
20	0.6870	1.3253	1.7247	2.0860	2.5280	2.8453
21	0.6864	1.3232	1.7207	2.0796	2.5177	2.8314
22	0.6858	1.3212	1.7171	2.0739	2.5083	2.8188
23	0.6853	1.3195	1.7139	2.0687	2.4999	2.8073
24	0.6848	1.3178	1.7109	2.0639	2.4922	2.7969
25	0.6844	1.3163	1.7081	2.0595	2.4851	2.7874
26	0.6840	1.3150	1.7056	2.0555	2.4786	2.7787

TABLE C.2   317

			Upper-Tail Areas			
Degrees of Freedom	0.25	0.10	0.05	0.025	0.01	0.005
27	0.6837	1.3137	1.7033	2.0518	2.4727	2.7707
28	0.6834	1.3125	1.7011	2.0484	2.4671	2.7633
29	0.6830	1.3114	1.6991	2.0452	2.4620	2.7564
30	0.6828	1.3104	1.6973	2.0423	2.4573	2.7500
31	0.6825	1.3095	1.6955	2.0395	2.4528	2.7440
32	0.6822	1.3086	1.6939	2.0369	2.4487	2.7385
33	0.6820	1.3077	1.6924	2.0345	2.4448	2.7333
34	0.6818	1.3070	1.6909	2.0322	2.4411	2.7284
35	0.6816	1.3062	1.6896	2.0301	2.4377	2.7238
36	0.6814	1.3055	1.6883	2.0281	2.4345	2.7195
37	0.6812	1.3049	1.6871	2.0262	2.4314	2.7154
38	0.6810	1.3042	1.6860	2.0244	2.4286	2.7116
39	0.6808	1.3036	1.6849	2.0227	2.4258	2.7079
40	0.6807	1.3031	1.6839	2.0211	2.4233	2.7045
41	0.6805	1.3025	1.6829	2.0195	2.4208	2.7012
42	0.6804	1.3020	1.6820	2.0181	2.4185	2.6981
43	0.6802	1.3016	1.6811	2.0167	2.4163	2.6951
44	0.6801	1.3011	1.6802	2.0154	2.4141	2.6923
45	0.6800	1.3006	1.6794	2.0141	2.4121	2.6896
46	0.6799	1.3022	1.6787	2.0129	2.4102	2.6870
47	0.6797	1.2998	1.6779	2.0117	2.4083	2.6846
48	0.6796	1.2994	1.6772	2.0106	2.4066	2.6822
49	0.6795	1.2991	1.6766	2.0096	2.4049	2.6800
50	0.6794	1.2987	1.6759	2.0086	2.4033	2.6778
51	0.6793	1.2984	1.6753	2.0076	2.4017	2.6757

(*continues*)

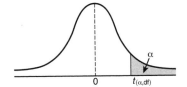

TABLE C.2 Continued

Degrees of Freedom	Upper-Tail Areas					
	0.25	0.10	0.05	0.025	0.01	0.005
52	0.6792	1.2980	1.6747	2.0066	2.4002	2.6737
53	0.6791	1.2977	1.6741	2.0057	2.3988	2.6718
54	0.6791	1.2974	1.6736	2.0049	2.3974	2.6700
55	0.6790	1.2971	1.6730	2.0040	2.3961	2.6682
56	0.6789	1.2969	1.6725	2.0032	2.3948	2.6665
57	0.6788	1.2966	1.6720	2.0025	2.3936	2.6649
58	0.6787	1.2963	1.6716	2.0017	2.3924	2.6633
59	0.6787	1.2961	1.6711	2.0010	2.3912	2.6618
60	0.6786	1.2958	1.6706	2.0003	2.3901	2.6603
61	0.6785	1.2956	1.6702	1.9996	2.3890	2.6589
62	0.6785	1.2954	1.6698	1.9990	2.3880	2.6575
63	0.6784	1.2951	1.6694	1.9983	2.3870	2.6561
64	0.6783	1.2949	1.6690	1.9977	2.3860	2.6549
65	0.6783	1.2947	1.6686	1.9971	2.3851	2.6536
66	0.6782	1.2945	1.6683	1.9966	2.3842	2.6524
67	0.6782	1.2943	1.6679	1.9960	2.3833	2.6512
68	0.6781	1.2941	1.6676	1.9955	2.3824	2.6501
69	0.6781	1.2939	1.6672	1.9949	2.3816	2.6490
70	0.6780	1.2938	1.6669	1.9944	2.3808	2.6479
71	0.6780	1.2936	1.6666	1.9939	2.3800	2.6469
72	0.6779	1.2934	1.6663	1.9935	2.3793	2.6459
73	0.6779	1.2933	1.6660	1.9930	2.3785	2.6449
74	0.6778	1.2931	1.6657	1.9925	2.3778	2.6439
75	0.6778	1.2929	1.6654	1.9921	2.3771	2.6430
76	0.6777	1.2928	1.6652	1.9917	2.3764	2.6421
77	0.6777	1.2926	1.6649	1.9913	2.3758	2.6412

TABLE C.2   319

Degrees of Freedom	Upper-Tail Areas					
	0.25	0.10	0.05	0.025	0.01	0.005
78	0.6776	1.2925	1.6646	1.9908	2.3751	2.6403
79	0.6776	1.2924	1.6644	1.9905	2.3745	2.6395
80	0.6776	1.2922	1.6641	1.9901	2.3739	2.6387
81	0.6775	1.2921	1.6639	1.9897	2.3733	2.6379
82	0.6775	1.2920	1.6636	1.9893	2.3727	2.6371
83	0.6775	1.2918	1.6634	1.9890	2.3721	2.6364
84	0.6774	1.2917	1.6632	1.9886	2.3716	2.6356
85	0.6774	1.2916	1.6630	1.9883	2.3710	2.6349
86	0.6774	1.2915	1.6628	1.9879	2.3705	2.6342
87	0.6773	1.2914	1.6626	1.9876	2.3700	2.6335
88	0.6773	1.2912	1.6624	1.9873	2.3695	2.6329
89	0.6773	1.2911	1.6622	1.9870	2.3690	2.6322
90	0.6772	1.2910	1.6620	1.9867	2.3685	2.6316
91	0.6772	1.2909	1.6618	1.9864	2.3680	2.6309
92	0.6772	1.2908	1.6616	1.9861	2.3676	2.6303
93	0.6771	1.2907	1.6614	1.9858	2.3671	2.6297
94	0.6771	1.2906	1.6612	1.9855	2.3667	2.6291
95	0.6771	1.2905	1.6611	1.9853	2.3662	2.6286
96	0.6771	1.2904	1.6609	1.9850	2.3658	2.6280
97	0.6770	1.2903	1.6607	1.9847	2.3654	2.6275
98	0.6770	1.2902	1.6606	1.9845	2.3650	2.6269
99	0.6770	1.2902	1.6604	1.9842	2.3646	2.6264
100	0.6770	1.2901	1.6602	1.9840	2.3642	2.6259
110	0.6767	1.2893	1.6588	1.9818	2.3607	2.6213
120	0.6765	1.2886	1.6577	1.9799	2.3578	2.6174
∞	0.6745	1.2816	1.6449	1.9600	2.3263	2.5758

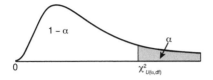

TABLE C.3

Critical Values of χ^2

For a particular number of degrees of freedom, entry represents the critical value of χ^2 corresponding to a specified upper-tail ar

Degrees of Freedom	Upper Tail Areas (α)					
	0.995	0.99	0.975	0.95	0.90	0.75
1	0.001	0.004	0.016	0.102	1.323	2.706
2	0.010	0.020	0.051	0.103	0.211	0.575
3	0.072	0.115	0.216	0.352	0.584	1.213
4	0.207	0.297	0.484	0.711	1.064	1.923
5	0.412	0.554	0.831	1.145	1.610	2.675
6	0.676	0.872	1.237	1.635	2.204	3.455
7	0.989	1.239	1.690	2.167	2.833	4.255
8	1.344	1.646	2.180	2.733	3.490	5.071
9	1.735	2.088	2.700	3.325	4.168	5.899
10	2.156	2.558	3.247	3.940	4.865	6.737
11	2.603	3.053	3.816	4.575	5.578	7.584
12	3.074	3.571	4.404	5.226	6.304	8.438
13	3.565	4.107	5.009	5.892	7.042	9.299
14	4.075	4.660	5.629	6.571	7.790	10.165
15	4.601	5.229	6.262	7.261	8.547	11.037
16	5.142	5.812	6.908	7.962	9.312	11.912
17	5.697	6.408	7.564	8.672	10.085	12.792
18	6.265	7.015	8.231	9.390	10.865	13.675
19	6.844	7.633	8.907	10.117	11.651	14.562
20	7.434	8.260	9.591	10.851	12.443	15.452
21	8.034	8.897	10.283	11.591	13.240	16.344
22	8.643	9.542	10.982	12.338	14.042	17.240
23	9.260	10.196	11.689	13.091	14.848	18.137
24	9.886	10.856	12.401	13.848	15.659	19.037
25	10.520	11.524	13.120	14.611	16.473	19.939
26	11.160	12.198	13.844	15.379	17.292	20.843
27	11.808	12.879	14.573	16.151	18.114	21.749
28	12.461	13.565	15.308	16.928	18.939	22.657
29	13.121	14.257	16.047	17.708	19.768	23.567
30	13.787	14.954	16.791	18.493	20.599	24.478

For larger values of degrees of freedom (df) the expression $Z = \sqrt{2\chi^2} - \sqrt{2(df)-1}$ may be used and the resulting upper-tail area can be obtained from the cumulative standardized normal distribution (Table C.1).

TABLE C.3   321

Upper Tail Areas (α)					
0.25	0.10	0.05	0.025	0.01	0.005
3.841	5.024	6.635	7.879		
2.773	4.605	5.991	7.378	9.210	10.597
4.108	6.251	7.815	9.348	11.345	12.838
5.385	7.779	9.488	11.143	13.277	14.860
6.626	9.236	11.071	12.833	15.086	16.750
7.841	10.645	12.592	14.449	16.812	18.458
9.037	12.017	14.067	16.013	18.475	20.278
10.219	13.362	15.507	17.535	20.090	21.955
11.389	14.684	16.919	19.023	21.666	23.589
12.549	15.987	18.307	20.483	23.209	25.188
13.701	17.275	19.675	21.920	24.725	26.757
14.845	18.549	21.026	23.337	26.217	28.299
15.984	19.812	22.362	24.736	27.688	29.819
17.117	21.064	23.685	26.119	29.141	31.319
18.245	22.307	24.996	27.488	30.578	32.801
19.369	23.542	26.296	28.845	32.000	34.267
20.489	24.769	27.587	30.191	33.409	35.718
21.605	25.989	28.869	31.526	34.805	37.156
22.718	27.204	30.144	32.852	36.191	38.582
23.828	28.412	31.410	34.170	37.566	39.997
24.935	29.615	32.671	35.479	38.932	41.401
26.039	30.813	33.924	36.781	40.289	42.796
27.141	32.007	35.172	38.076	41.638	44.181
28.241	33.196	36.415	39.364	42.980	45.559
29.339	34.382	37.652	40.646	44.314	46.928
30.435	35.563	38.885	41.923	45.642	48.290
31.528	36.741	40.113	43.194	46.963	49.645
32.620	37.916	41.337	44.461	48.278	50.993
33.711	39.087	42.557	45.722	49.588	52.336
34.800	40.256	43.773	46.979	50.892	53.672

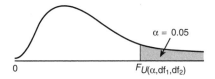

TABLE C.4

Critical Values of F

For a particular combination of numerator and denominator degrees of freedom, entry represents the critical values of F corresponding to a specified upper-tail area (α).

Denominator, df_2	Numerator, df_1 1	2	3	4	5	6	7	8	9
1	161.40	199.50	215.70	224.60	230.20	234.00	236.80	238.90	240.50
2	18.51	19.00	19.16	19.25	19.30	19.33	19.35	19.37	19.38
3	10.13	9.55	9.28	9.12	9.01	8.94	8.89	8.85	8.81
4	7.71	6.94	6.59	6.39	6.26	6.16	6.09	6.04	6.00
5	6.61	5.79	5.41	5.19	5.05	4.95	4.88	4.82	4.77
6	5.99	5.14	4.76	4.53	4.39	4.28	4.21	4.15	4.10
7	5.59	4.74	4.35	4.12	3.97	3.87	3.79	3.73	3.68
8	5.32	4.46	4.07	3.84	3.69	3.58	3.50	3.44	3.39
9	5.12	4.26	3.86	3.63	3.48	3.37	3.29	3.23	3.18
10	4.96	4.10	3.71	3.48	3.33	3.22	3.14	3.07	3.02
11	4.84	3.98	3.59	3.36	3.20	3.09	3.01	2.95	2.90
12	4.75	3.89	3.49	3.26	3.11	3.00	2.91	2.85	2.80
13	4.67	3.81	3.41	3.18	3.03	2.92	2.83	2.77	2.71
14	4.60	3.74	3.34	3.11	2.96	2.85	2.76	2.70	2.65
15	4.54	3.68	3.29	3.06	2.90	2.79	2.71	2.64	2.59
16	4.49	3.63	3.24	3.01	2.85	2.74	2.66	2.59	2.54
17	4.45	3.59	3.20	2.96	2.81	2.70	2.61	2.55	2.49
18	4.41	3.55	3.16	2.93	2.77	2.66	2.58	2.51	2.46
19	4.38	3.52	3.13	2.90	2.74	2.63	2.54	2.48	2.42

TABLE C.4   323

				Numerator, df$_1$					
10	**12**	**15**	**20**	**24**	**30**	**40**	**60**	**120**	**∞**
241.90	243.90	245.90	248.00	249.10	250.10	251.10	252.20	253.30	254.30
19.40	19.41	19.43	19.45	19.45	19.46	19.47	19.48	19.49	19.50
8.79	8.74	8.70	8.66	8.64	8.62	8.59	8.57	8.55	8.53
5.96	5.91	5.86	5.80	5.77	5.75	5.72	5.69	5.66	5.63
4.74	4.68	4.62	4.56	4.53	4.50	4.46	4.43	4.40	4.36
4.06	4.00	3.94	3.87	3.84	3.81	3.77	3.74	3.70	3.67
3.64	3.57	3.51	3.44	3.41	3.38	3.34	3.30	3.27	3.23
3.35	3.28	3.22	3.15	3.12	3.08	3.04	3.01	2.97	2.93
3.14	3.07	3.01	2.94	2.90	2.86	2.83	2.79	2.75	2.71
2.98	2.91	2.85	2.77	2.74	2.70	2.66	2.62	2.58	2.54
2.85	2.79	2.72	2.65	2.61	2.57	2.53	2.49	2.45	2.40
2.75	2.69	2.62	2.54	2.51	2.47	2.43	2.38	2.34	2.30
2.67	2.60	2.53	2.46	2.42	2.38	2.34	2.30	2.25	2.21
2.60	2.53	2.46	2.39	2.35	2.31	2.27	2.22	2.18	2.13
2.54	2.48	2.40	2.33	2.29	2.25	2.20	2.16	2.11	2.07
2.49	2.42	2.35	2.28	2.24	2.19	2.15	2.11	2.06	2.01
2.45	2.38	2.31	2.23	2.19	2.15	2.10	2.06	2.01	1.96
2.41	2.34	2.27	2.19	2.15	2.11	2.06	2.02	1.97	1.92
2.38	2.31	2.23	2.16	2.11	2.07	2.03	1.98	1.93	1.88

(continues)

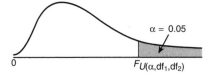

TABLE C.4 Continued

For a particular combination of numerator and denominator degrees of freedom, entry represents the critical values of F corresponding to a specified upper-tail area (α).

Denominator, df_2	Numerator, df_1								
	1	2	3	4	5	6	7	8	9
20	4.35	3.49	3.10	2.87	2.71	2.60	2.51	2.45	2.39
21	4.32	3.47	3.07	2.84	2.68	2.57	2.49	2.42	2.37
22	4.30	3.44	3.05	2.82	2.66	2.55	2.46	2.40	2.34
23	4.28	3.42	3.03	2.80	2.64	2.53	2.44	2.37	2.32
24	4.26	3.40	3.01	2.78	2.62	2.51	2.42	2.36	2.30
25	4.24	3.39	2.99	2.76	2.60	2.49	2.40	2.34	2.28
26	4.23	3.37	2.98	2.74	2.59	2.47	2.39	2.32	2.27
27	4.21	3.35	2.96	2.73	2.57	2.46	2.37	2.31	2.25
28	4.20	3.34	2.95	2.71	2.56	2.45	2.36	2.29	2.24
29	4.18	3.33	2.93	2.70	2.55	2.43	2.35	2.28	2.22
30	4.17	3.32	2.92	2.69	2.53	2.42	2.33	2.27	2.21
40	4.08	3.23	2.84	2.61	2.45	2.34	2.25	2.18	2.12
60	4.00	3.15	2.76	2.53	2.37	2.25	2.17	2.10	2.04
120	3.92	3.07	2.68	2.45	2.29	2.17	2.09	2.02	1.96
∞	3.84	3.00	2.60	2.37	2.21	2.10	2.01	1.94	1.88

TABLE C.4   325

			Numerator, df$_1$						
10	**12**	**15**	**20**	**24**	**30**	**40**	**60**	**120**	**∞**
2.35	2.28	2.20	2.12	2.08	2.04	1.99	1.95	1.90	1.84
2.32	2.25	2.18	2.10	2.05	2.01	1.96	1.92	1.87	1.81
2.30	2.23	2.15	2.07	2.03	1.98	1.91	1.89	1.84	1.78
2.27	2.20	2.13	2.05	2.01	1.96	1.91	1.86	1.81	1.76
2.25	2.18	2.11	2.03	1.98	1.94	1.89	1.84	1.79	1.73
2.24	2.16	2.09	2.01	1.96	1.92	1.87	1.82	1.77	1.71
2.22	2.15	2.07	1.99	1.95	1.90	1.85	1.80	1.75	1.69
2.20	2.13	2.06	1.97	1.93	1.88	1.84	1.79	1.73	1.67
2.19	2.12	2.04	1.96	1.91	1.87	1.82	1.77	1.71	1.65
2.18	2.10	2.03	1.94	1.90	1.85	1.81	1.75	1.70	1.64
2.16	2.09	2.01	1.93	1.89	1.84	1.79	1.74	1.68	1.62
2.08	2.00	1.92	1.84	1.79	1.74	1.69	1.64	1.58	1.51
1.99	1.92	1.84	1.75	1.70	1.65	1.59	1.53	1.47	1.39
1.91	1.83	1.75	1.66	1.61	1.55	1.50	1.43	1.35	1.25
1.83	1.75	1.67	1.57	1.52	1.46	1.39	1.32	1.22	1.00

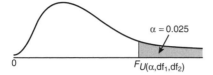

TABLE C.4 Continued

For a particular combination of numerator and denominator degrees of freedom, entry represents the critical values of F corresponding to a specified upper-tail area (α).

Denominator, df$_2$	Numerator, df$_1$								
	1	2	3	4	5	6	7	8	9
1	647.80	799.50	864.20	899.60	921.80	937.10	948.20	956.70	963.30
2	38.51	39.00	39.17	39.25	39.30	39.33	39.36	39.39	39.39
3	17.44	16.04	15.44	15.10	14.88	14.73	14.62	14.54	14.47
4	12.22	10.65	9.98	9.60	9.36	9.20	9.07	8.98	8.90
5	10.01	8.43	7.76	7.39	7.15	6.98	6.85	6.76	6.68
6	8.81	7.26	6.60	6.23	5.99	5.82	5.70	5.60	5.52
7	8.07	6.54	5.89	5.52	5.29	5.12	4.99	4.90	4.82
8	7.57	6.06	5.42	5.05	4.82	4.65	4.53	4.43	4.36
9	7.21	5.71	5.08	4.72	4.48	4.32	4.20	4.10	4.03
10	6.94	5.46	4.83	4.47	4.24	4.07	3.95	3.85	3.78
11	6.72	5.26	4.63	4.28	4.04	3.88	3.76	3.66	3.59
12	6.55	5.10	4.47	4.12	3.89	3.73	3.61	3.51	3.44
13	6.41	4.97	4.35	4.00	3.77	3.60	3.48	3.39	3.31
14	6.30	4.86	4.24	3.89	3.66	3.50	3.38	3.29	3.21
15	6.20	4.77	4.15	3.80	3.58	3.41	3.29	3.20	3.12
16	6.12	4.69	4.08	3.73	3.50	3.34	3.22	3.12	3.05
17	6.04	4.62	4.01	3.66	3.44	3.28	3.16	3.06	2.98
18	5.98	4.56	3.95	3.61	3.38	3.22	3.10	3.01	2.93
19	5.92	4.51	3.90	3.56	3.33	3.17	3.05	2.96	2.88
20	5.87	4.46	3.86	3.51	3.29	3.13	3.01	2.91	2.84
21	5.83	4.42	3.82	3.48	3.25	3.09	2.97	2.87	2.80

TABLE C.4   327

				Numerator, df$_1$					
10	**12**	**15**	**20**	**24**	**30**	**40**	**60**	**120**	**∞**
968.60	976.70	984.90	993.10	997.20	1,001.00	1,006.00	1,010.00	1,014.00	1,018.00
39.40	39.41	39.43	39.45	39.46	39.46	39.47	39.48	39.49	39.50
14.42	14.34	14.25	14.17	14.12	14.08	14.04	13.99	13.95	13.90
8.84	8.75	8.66	8.56	8.51	8.46	8.41	8.36	8.31	8.26
6.62	6.52	6.43	6.33	6.28	6.23	6.18	6.12	6.07	6.02
5.46	5.37	5.27	5.17	5.12	5.07	5.01	4.96	4.90	4.85
4.76	4.67	4.57	4.47	4.42	4.36	4.31	4.25	4.20	4.14
4.30	4.20	4.10	4.00	3.95	3.89	3.84	3.78	3.73	3.67
3.96	3.87	3.77	3.67	3.61	3.56	3.51	3.45	3.39	3.33
3.72	3.62	3.52	3.42	3.37	3.31	3.26	3.20	3.14	3.08
3.53	3.43	3.33	3.23	3.17	3.12	3.06	3.00	2.94	2.88
3.37	3.28	3.18	3.07	3.02	2.96	2.91	2.85	2.79	2.72
3.25	3.15	3.05	2.95	2.89	2.84	2.78	2.72	2.66	2.60
3.15	3.05	2.95	2.84	2.79	2.73	2.67	2.61	2.55	2.49
3.06	2.96	2.86	2.76	2.70	2.64	2.59	2.52	2.46	2.40
2.99	2.89	2.79	2.68	2.63	2.57	2.51	2.45	2.38	2.32
2.92	2.82	2.72	2.62	2.56	2.50	2.44	2.38	2.32	2.25
2.87	2.77	2.67	2.56	2.50	2.44	2.38	2.32	2.26	2.19
2.82	2.72	2.62	2.51	2.45	2.39	2.33	2.27	2.20	2.13
2.77	2.68	2.57	2.46	2.41	2.35	2.29	2.22	2.16	2.09
2.73	2.64	2.53	2.42	2.37	2.31	2.25	2.18	2.11	2.04

(*continues*)

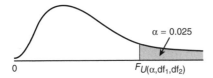

TABLE C.4 Continued

For a particular combination of numerator and denominator degrees of freedom, entry represents the critical values of F corresponding to a specified upper-tail area (α).

Denominator, df₂	Numerator, df₁								
	1	2	3	4	5	6	7	8	9
22	5.79	4.38	3.78	3.44	3.22	3.05	2.93	2.84	2.76
23	5.75	4.35	3.75	3.41	3.18	3.02	2.90	2.81	2.73
24	5.72	4.32	3.72	3.38	3.15	2.99	2.87	2.78	2.70
25	5.69	4.29	3.69	3.35	3.13	2.97	2.85	2.75	2.68
26	5.66	4.27	3.67	3.33	3.10	2.94	2.82	2.73	2.65
27	5.63	4.24	3.65	3.31	3.08	2.92	2.80	2.71	2.63
28	5.61	4.22	3.63	3.29	3.06	2.90	2.78	2.69	2.61
29	5.59	4.20	3.61	3.27	3.04	2.88	2.76	2.67	2.59
30	5.57	4.18	3.59	3.25	3.03	2.87	2.75	2.65	2.57
40	5.42	4.05	3.46	3.13	2.90	2.74	2.62	2.53	2.45
60	5.29	3.93	3.34	3.01	2.79	2.63	2.51	2.41	2.33
120	5.15	3.80	3.23	2.89	2.67	2.52	2.39	2.30	2.22
∞	5.02	3.69	3.12	2.79	2.57	2.41	2.29	2.19	2.11

TABLE C.4   329

			Numerator, df$_1$						
10	12	15	20	24	30	40	60	120	∞
2.70	2.60	2.50	2.39	2.33	2.27	2.21	2.14	2.08	2.00
2.67	2.57	2.47	2.36	2.30	2.24	2.18	2.11	2.04	1.97
2.64	2.54	2.44	2.33	2.27	2.21	2.15	2.08	2.01	1.94
2.61	2.51	2.41	2.30	2.24	2.18	2.12	2.05	1.98	1.91
2.59	2.49	2.39	2.28	2.22	2.16	2.09	2.03	1.95	1.88
2.57	2.47	2.36	2.25	2.19	2.13	2.07	2.00	1.93	1.85
2.55	2.45	2.34	2.23	2.17	2.11	2.05	1.98	1.91	1.83
2.53	2.43	2.32	2.21	2.15	2.09	2.03	1.96	1.89	1.81
2.51	2.41	2.31	2.20	2.14	2.07	2.01	1.94	1.87	1.79
2.39	2.29	2.18	2.07	2.01	1.94	1.88	1.80	1.72	1.64
2.27	2.17	2.06	1.94	1.88	1.82	1.74	1.67	1.58	1.48
2.16	2.05	1.94	1.82	1.76	1.69	1.61	1.53	1.43	1.31
2.05	1.94	1.83	1.71	1.64	1.57	1.48	1.39	1.27	1.00

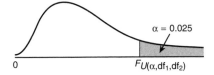

$\alpha = 0.025$

$F_{U(\alpha, df_1, df_2)}$

TABLE C.4 Continued

For a particular combination of numerator and denominator degrees of freedom, entry represents the critical values of F corresponding to a specified upper-tail area (α).

Denominator, df_2	Numerator, df_1								
	1	2	3	4	5	6	7	8	9
1	4,052.00	4,999.50	5,403.00	5,625.00	5,764.00	5,859.00	5,928.00	5,982.00	6,022.00
2	98.50	99.00	99.17	99.25	99.30	99.33	99.36	99.37	99.39
3	34.12	30.82	29.46	28.71	28.24	27.91	27.67	27.49	27.35
4	21.20	18.00	16.69	15.98	15.52	15.21	14.98	14.80	14.66
5	16.26	13.27	12.06	11.39	10.97	10.67	10.46	10.29	10.16
6	13.75	10.92	9.78	9.15	8.75	8.47	8.26	8.10	7.98
7	12.25	9.55	8.45	7.85	7.46	7.19	6.99	6.84	6.72
8	11.26	8.65	7.59	7.01	6.63	6.37	6.18	6.03	5.91
9	10.56	8.02	6.99	6.42	6.06	5.80	5.61	5.47	5.35
10	10.04	7.56	6.55	5.99	5.64	5.39	5.20	5.06	4.94
11	9.65	7.21	6.22	5.67	5.32	5.07	4.89	4.74	4.63
12	9.33	6.93	5.95	5.41	5.06	4.82	4.64	4.50	4.39
13	9.07	6.70	5.74	5.21	4.86	4.62	4.44	4.30	4.19
14	8.86	6.51	5.56	5.04	4.69	4.46	4.28	4.14	4.03
15	8.68	6.36	5.42	4.89	4.56	4.32	4.14	4.00	3.89
16	8.53	6.23	5.29	4.77	4.44	4.20	4.03	3.89	3.78
17	8.40	6.11	5.18	4.67	4.34	4.10	3.93	3.79	3.68
18	8.29	6.01	5.09	4.58	4.25	4.01	3.84	3.71	3.60
19	8.18	5.93	5.01	4.50	4.17	3.94	3.77	3.63	3.52
20	8.10	5.85	4.94	4.43	4.10	3.87	3.70	3.56	3.46
21	8.02	5.78	4.87	4.37	4.04	3.81	3.64	3.51	3.40

TABLE C.4   331

				Numerator, df$_1$					
10	**12**	**15**	**20**	**24**	**30**	**40**	**60**	**120**	**∞**
6,056.00	6,106.00	6,157.00	6,209.00	6,235.00	6,261.00	6,287.00	6,313.00	6,339.00	6,366.00
99.40	99.42	99.43	94.45	99.46	99.47	99.47	99.48	99.49	99.50
27.23	27.05	26.87	26.69	26.60	26.50	26.41	26.32	26.22	26.13
14.55	14.37	14.20	14.02	13.93	13.84	13.75	13.65	13.56	13.46
10.05	9.89	9.72	9.55	9.47	9.38	9.29	9.20	9.11	9.02
7.87	7.72	7.56	7.40	7.31	7.23	7.14	7.06	6.97	6.88
6.62	6.47	6.31	6.16	6.07	5.99	5.91	5.82	5.74	5.65
5.81	5.67	5.52	5.36	5.28	5.20	5.12	5.03	4.95	4.86
5.26	5.11	4.96	4.81	4.73	4.65	4.57	4.48	4.40	4.31
4.85	4.71	4.56	4.41	4.33	4.25	4.17	4.08	4.00	3.91
4.54	4.40	4.25	4.10	4.02	3.94	3.86	3.78	3.69	3.60
4.30	4.16	4.01	3.86	3.78	3.70	3.62	3.54	3.45	3.36
4.10	3.96	3.82	3.66	3.59	3.51	3.43	3.34	3.25	3.17
3.94	3.80	3.66	3.51	3.43	3.35	3.27	3.18	3.09	3.00
3.80	3.67	3.52	3.37	3.29	3.21	3.13	3.05	2.96	2.87
3.69	3.55	3.41	3.26	3.18	3.10	3.02	2.93	2.81	2.75
3.59	3.46	3.31	3.16	3.08	3.00	2.92	2.83	2.75	2.65
3.51	3.37	3.23	3.08	3.00	2.92	2.84	2.75	2.66	2.57
3.43	3.30	3.15	3.00	2.92	2.84	2.76	2.67	2.58	2.49
3.37	3.23	3.09	2.94	2.86	2.78	2.69	2.61	2.52	2.42
3.31	3.17	3.03	2.88	2.80	2.72	2.64	2.55	2.46	2.36

(*continues*)

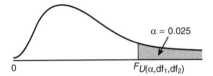

TABLE C.4 Continued

For a particular combination of numerator and denominator degrees of freedom, entry represents the critical values of F corresponding to a specified upper-tail area (α).

Denominator, df$_2$	\	\	\	Numerator, df$_1$	\	\	\	\	\
	1	2	3	4	5	6	7	8	9
22	7.95	5.72	4.82	4.31	3.99	3.76	3.59	3.45	3.35
23	7.88	5.66	4.76	4.26	3.94	3.71	3.54	3.41	3.30
24	7.82	5.61	4.72	4.22	3.90	3.67	3.50	3.36	3.26
25	7.77	5.57	4.68	4.18	3.85	3.63	3.46	3.32	3.22
26	7.72	5.53	4.64	4.14	3.82	3.59	3.42	3.29	3.18
27	7.68	5.49	4.60	4.11	3.78	3.56	3.39	3.26	3.15
28	7.64	5.45	4.57	4.07	3.75	3.53	3.36	3.23	3.12
29	7.60	5.42	4.54	4.04	3.73	3.50	3.33	3.20	3.09
30	7.56	5.39	4.51	4.02	3.70	3.47	3.30	3.17	3.07
40	7.31	5.18	4.31	3.83	3.51	3.29	3.12	2.99	2.89
60	7.08	4.98	4.13	3.65	3.34	3.12	2.95	2.82	2.72
120	6.85	4.79	3.95	3.48	3.17	2.96	2.79	2.66	2.56
∞	6.63	4.61	3.78	3.32	3.02	2.80	2.64	2.51	2.41

TABLE C.4   333

					Numerator, df_1				
10	**12**	**15**	**20**	**24**	**30**	**40**	**60**	**120**	**∞**
3.26	3.12	2.98	2.83	2.75	2.67	2.58	2.50	2.40	2.31
3.21	3.07	2.93	2.78	2.70	2.62	2.54	2.45	2.35	2.26
3.17	3.03	2.89	2.74	2.66	2.58	2.49	2.40	2.31	2.21
3.13	2.99	2.85	2.70	2.62	2.54	2.45	2.36	2.27	2.17
3.09	2.96	2.81	2.66	2.58	2.50	2.42	2.33	2.23	2.13
3.06	2.93	2.78	2.63	2.55	2.47	2.38	2.29	2.20	2.10
3.03	2.90	2.75	2.60	2.52	2.44	2.35	2.26	2.17	2.06
3.00	2.87	2.73	2.57	2.49	2.41	2.33	2.23	2.14	2.03
2.98	2.84	2.70	2.55	2.47	2.39	2.30	2.21	2.11	2.01
2.80	2.66	2.52	2.37	2.29	2.20	2.11	2.02	1.92	1.80
2.63	2.50	2.35	2.20	2.12	2.03	1.94	1.84	1.73	1.60
2.47	2.34	2.19	2.03	1.95	1.86	1.76	1.66	1.53	1.38
2.32	2.18	2.04	1.88	1.79	1.70	1.59	1.47	1.32	1.00

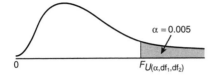

TABLE C.4 Continued

Denominator, df_2	Numerator, df_1								
	1	**2**	**3**	**4**	**5**	**6**	**7**	**8**	
1	16,211.00	20,000.000	21,615.00	22,500.00	23,056.00	23,437.00	23,715.00	23,925.00	24,(
2	198.50	199.00	199.20	199.20	199.30	199.30	199.40	199.40	1
3	55.55	49.80	47.47	46.19	45.39	44.84	44.43	44.13	
4	31.33	26.28	24.26	23.15	22.46	21.97	21.62	21.35	
5	22.78	18.31	16.53	15.56	14.94	14.51	14.20	13.96	
6	18.63	14.54	12.92	12.03	11.46	11.07	10.79	10.57	
7	16.24	12.40	10.88	10.05	9.52	9.16	8.89	8.68	
8	14.69	11.04	9.60	8.81	8.30	7.95	7.69	7.50	
9	13.61	10.11	8.72	7.96	7.47	7.13	6.88	6.69	
10	12.83	9.43	8.08	7.34	6.87	6.54	6.30	6.12	
11	12.23	8.91	7.60	6.88	6.42	6.10	5.86	5.68	
12	11.75	8.51	7.23	6.52	6.07	5.76	5.52	5.35	
13	11.37	8.19	6.93	6.23	5.79	5.48	5.25	5.08	
14	11.06	7.92	6.68	6.00	5.56	5.26	5.03	4.86	
15	10.80	7.70	6.48	5.80	5.37	5.07	4.85	4.67	
16	10.58	7.51	6.30	5.64	5.21	4.91	4.69	4.52	
17	10.38	7.35	6.16	5.50	5.07	4.78	4.56	4.39	
18	10.22	7.21	6.03	5.37	4.96	4.66	4.44	4.28	
19	10.07	7.09	5.92	5.27	4.85	4.56	4.34	4.18	
20	9.94	6.99	5.82	5.17	4.76	4.47	4.26	4.09	
21	9.83	6.89	5.73	5.09	4.68	4.39	4.18	4.02	
22	9.73	6.81	5.65	5.02	4.61	4.32	4.11	3.94	
23	9.63	6.73	5.58	4.95	4.54	4.26	4.05	3.88	
24	9.55	6.66	5.52	4.89	4.49	4.20	3.99	3.83	

TABLE C.4   335

			Numerator, df$_1$						
10	**12**	**15**	**20**	**24**	**30**	**40**	**60**	**120**	**∞**
24,224.00	24,426.00	24,630.00	24,836.00	24,910.00	25,044.00	25,148.00	25,253.00	25,359.00	25,465.00
199.40	199.40	199.40	199.40	199.50	199.50	199.50	199.50	199.50	199.50
43.69	43.39	43.08	42.78	42.62	42.47	42.31	42.15	41.99	41.83
20.97	20.70	20.44	20.17	20.03	19.89	19.75	19.61	19.47	19.32
13.62	13.38	13.15	12.90	12.78	12.66	12.53	12.40	12.27	12.11
10.25	10.03	9.81	9.59	9.47	9.36	9.24	9.12	9.00	8.88
8.38	8.18	7.97	7.75	7.65	7.53	7.42	7.31	7.19	7.08
7.21	7.01	6.81	6.61	6.50	6.40	6.29	6.18	6.06	5.95
6.42	6.23	6.03	5.83	5.73	5.62	5.52	5.41	5.30	5.19
5.85	5.66	5.47	5.27	5.17	5.07	4.97	4.86	4.75	1.61
5.42	5.24	5.05	4.86	4.75	4.65	4.55	4.44	4.34	4.23
5.09	4.91	4.72	4.53	4.43	4.33	4.23	4.12	4.01	3.90
4.82	4.64	4.46	4.27	4.17	4.07	3.97	3.87	3.76	3.65
4.60	4.43	4.25	4.06	3.96	3.86	3.76	3.66	3.55	3.41
4.42	4.25	4.07	3.88	3.79	3.69	3.58	3.48	3.37	3.26
4.27	4.10	3.92	3.73	3.64	3.54	3.44	3.33	3.22	3.11
4.14	3.97	3.79	3.61	3.51	3.41	3.31	3.21	3.10	2.98
4.03	3.86	3.68	3.50	3.40	3.30	3.20	3.10	2.89	2.87
3.93	3.76	3.59	3.40	3.31	3.21	3.11	3.00	2.89	2.78
3.85	3.68	3.50	3.32	3.22	3.12	3.02	2.92	2.81	2.69
3.77	3.60	3.43	3.24	3.15	3.05	2.95	2.84	2.73	2.61
3.70	3.54	3.36	3.18	3.08	2.98	2.88	2.77	2.66	2.55
3.64	3.47	3.30	3.12	3.02	2.92	2.82	2.71	2.60	2.48
3.59	3.42	3.25	3.06	2.97	2.87	2.77	2.66	2.55	2.43

(continues)

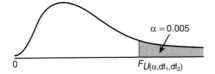

TABLE C.4 Continued

Denominator, df$_2$	Numerator, df$_1$							
	1	**2**	**3**	**4**	**5**	**6**	**7**	**8**
25	9.48	6.60	5.46	4.84	4.43	4.15	3.94	3.78
26	9.41	6.54	5.41	4.79	4.38	4.10	3.89	3.73
27	9.34	6.49	5.36	4.74	4.34	4.06	3.85	3.69
28	9.28	6.44	5.32	4.70	4.30	4.02	3.81	3.65
29	9.23	6.40	5.28	4.66	4.26	3.98	3.77	3.61
30	9.18	6.35	5.24	4.62	4.23	3.95	3.74	3.58
40	8.83	6.07	4.98	4.37	3.99	3.71	3.51	3.35
60	8.49	5.79	4.73	4.14	3.76	3.49	3.29	3.13
120	8.18	5.54	4.50	3.92	3.55	3.28	3.09	2.93
∞	7.88	5.30	4.28	3.72	3.35	3.09	2.90	2.74

TABLE C.4 337

				Numerator, df$_1$					
10	12	15	20	24	30	40	60	120	∞
3.54	3.37	3.20	3.01	2.92	2.82	2.72	2.61	2.50	2.38
3.49	3.33	3.15	2.97	2.87	2.77	2.67	2.56	2.45	2.33
3.45	3.28	3.11	2.93	2.83	2.73	2.63	2.52	2.41	2.29
3.41	3.25	3.07	2.89	2.79	2.69	2.59	2.48	2.37	2.25
3.38	3.21	3.04	2.86	2.76	2.66	2.56	2.45	2.33	2.21
3.34	3.18	3.01	2.82	2.73	2.63	2.52	2.42	2.30	2.18
3.12	2.95	2.78	2.60	2.50	2.40	2.30	2.18	2.06	1.93
2.90	2.74	2.57	2.39	2.29	2.19	2.08	1.96	1.83	1.69
2.71	2.54	2.37	2.19	2.09	1.98	1.87	1.75	1.61	1.43
2.52	2.36	2.19	2.00	1.90	1.79	1.67	1.53	1.36	1.00

TABLE C.5

Control Chart Factors

Number of Observations in Sample	d_2	d_3	D_3	D_4	A_2
2	1.128	0.853	0	3.267	1.880
3	1.693	0.888	0	2.575	1.023
4	2.059	0.880	0	2.282	0.729
5	2.326	0.864	0	2.114	0.577
6	2.534	0.848	0	2.004	0.483
7	2.704	0.833	0.076	1.924	0.419
8	2.847	0.820	0.136	1.864	0.373
9	2.970	0.808	0.184	1.816	0.337
10	3.078	0.797	0.223	1.777	0.308
11	3.173	0.787	0.256	1.744	0.285
12	3.258	0.778	0.283	1.717	0.266
13	3.336	0.770	0.307	1.693	0.249
14	3.407	0.763	0.328	1.672	0.235
15	3.472	0.756	0.347	1.653	0.223
16	3.532	0.750	0.363	1.637	0.212
17	3.588	0.744	0.378	1.622	0.203
18	3.640	0.739	0.391	1.609	0.194
19	3.689	0.733	0.404	1.596	0.187
20	3.735	0.729	0.415	1.585	0.180
21	3.778	0.724	0.425	1.575	0.173
22	3.819	0.720	0.435	1.565	0.167
23	3.858	0.716	0.443	1.557	0.162
24	3.895	0.712	0.452	1.548	0.157
25	3.931	0.708	0.459	1.541	0.153

Source: Reprinted from ASTM-STP 15D by kind permission of the American Society for Testing and Materials.

D

Spreadsheet Tips

These Spreadsheet Tips complement the Spreadsheet Solutions that appear throughout this book. Use these tips as a starting point to learn more about how to use Microsoft Excel or OpenOffice.org Calc 3 to obtain your own customized results. The tips are divided into three groups:

- **Chart Tips** that assist you in creating and refining your own charts.
- **Function Tips** that explain the worksheet statistical functions used in Spreadsheet Solutions
- **Analysis ToolPak Tips** that review how to use selected Analysis ToolPak statistical procedures (Microsoft Excel only)

CT: Chart Tips

CT1 Arranging Data in Categorical Charts

You can sort your data to display categories in largest to smallest order. For bar charts, arrange the summary table in largest to smallest order. For pie charts, arrange the summary table in largest to smallest order in order to show largest to smallest pie slices in clockwise order.

CT2 Reformatting Charts

Right-click on chart components to reformat chart components. Use this technique to eliminate unwarranted gridlines and legends, change color schemes, or to change the text font and size of titles and axis labels.

In OpenOffice.org Calc 3, right-click and initially click **Edit**. Then right-click on the chart component a second time to reformat the component.

CT3 Creating Charts

How you create a chart depends on whether you are using Excel 2007. If you use Excel 2007, you first select the data to be charted by dragging your mouse over the data and then you make a choice in the **Charts** group of the **Insert** tab (of the **Office Ribbon**). You can restyle charts using the choices on the **Chart Tools Design** or **Layout** tabs.

If you use an earlier version of Excel or use OpenOffice.org Calc 3, you select **Insert**, then **Chart** and then work through a **Chart Wizard**, a series of dialog boxes that ask you for the chart type, the cell range(s) of the data to be charted, and chart elements such as titles and axis labels that you want to use.

CT4 Creating Pareto Charts

To create a Pareto chart, first create a summary table with columns for categories, percentage, and cumulative percentage, arranged from largest to smallest percentage. Then create a new chart that uses this summary table. Select **Clustered Column** in Excel 2007, **Line – Column** in older Excel versions, or **Column and Line** in OpenOffice.org Calc 3. In Excel 2007, you must click on the second series (the cumulative percentages) and then click **Change Chart Type** in the **Chart Tools Design** tab to change that series from a column to line type.

Add data labels using your categories column. Adjust the primary (left) Y-axis so that it shows a scale from 0 to 100%. Add a secondary (right) Y-axis with the same scale. Reformat the chart as necessary using Tip CT2.

CT5 Creating Frequency Distributions

If you use Microsoft Excel, the Analysis ToolPak **Histogram** procedure is the easiest way to create a frequency distribution from your data values. If you use OpenOffice.org Calc 3, you must use the **FREQUENCY** function, discussed in section E.2 of Appendix E, "Advanced Techniques." (You can also use the **FREQUENCY** function in Microsoft Excel.)

In either case, you need to first create a column of bin values on the worksheet that contains your data values. Bin values represent the maximum value of a group. When groups are defined in the form *low value through high value*, the bin value is the *high value*. When groups are defined in the form *low value to under high value*, as they are in Chapter 2, "Presenting Data in Charts and Tables," the bin value is a number just smaller than the *high value*. For example, for the group "150 to under 200," the corresponding bin value could be 199.99, a number "just smaller" than 200.

With the column of bin values created, you continue by using the **Histogram** procedure (see Tip ATT1 later in this appendix) or by entering its cell range as part of the **FREQUENCY** function (see section E.2).

CT6 Creating Histograms

If you have previously chosen to use the Analysis ToolPak **Histogram** procedure, complete all the instructions in Tip ATT1 (later in this appendix) to create a histogram. If you have created a frequency distribution using the technique in section E.2 or have data already summarized into a frequency distribution, create a new **Column** chart (see Tip CT3, if necessary) to create a histogram. You also can create a column of midpoints to serve as the *X*-axis labels, as is done in Chapter 2's "Histogram" section.

CT7 Creating Scatter Plots

First, arrange your data values so that the values to be plotted on the *X* (horizontal) axis appear in the column to the immediate left of the values to be plotted on the *Y* axis. Then create a new chart and select **XY (Scatter)** as the chart type.

FT: Function Tips

FT1 Measures of Central Tendency

Use the **AVERAGE** (for the mean), **MEDIAN**, and **MODE** functions in worksheet formulas to calculate measures of central tendency. Enter these functions in the form FUNCTION(*cell range of the data values*).

FT2 Measures of Variation

Use the **VAR** (sample variance), **STDEV** (sample standard deviation), **VARP** (population variance), and **STDEVP** (population standard deviation) functions to calculate measures of variation. Use the difference of the **MAX** (maximum value) and **MIN** (minimum value) functions to calculate the range.

Enter these functions in the form FUNCTION(*cell range of the data values*).

FT3 Binomial Probabilities

Use the **BINOMDIST** function to calculate binomial probabilities. Enter the function as **BINOMDIST(X, n, p, *cumulative*)**, where X is the number of successes, n is the sample size, p is the probability of success, and *cumulative* is a **True** or **False** value in which True causes the function to calculate the probability of X or fewer successes and False, the probability of exactly X successes.

FT4 Poisson Probabilities

Use the **POISSON** function to calculate Poisson probabilities. Enter the function as **POISSON(X, *lambda*, *cumulative*)**, where X is the number of successes, *lambda* is the mean or expected number of successes, and *cumulative* is a **True** or **False** value in which True causes the function to calculate the probability of X or fewer successes and False, the probability of exactly X successes.

FT5 Normal Probabilities

Use the **STANDARDIZE, NORMDIST, NORMSINV**, and **NORMINV** functions to calculate values associated with normal probabilities. Enter these functions as

- **STANDARDIZE(X, *mean*, *standard deviation*)**, where X is the X value of interest, and *mean* and *standard deviation* are the mean and standard deviation for a variable of interest
- **NORMDIST(X, *mean*, *standard deviation*, *True*)**
- **NORMSINV(P<X)**, where $P<X$ is the area under the curve that is less than X
- **NORMINV(P<X, *mean*, *standard deviation*)**

The STANDARDIZE function returns the Z value for a particular X value, mean, and standard deviation. The NORMDIST function returns the area or probability of less than a given X value. The NORMSINV function returns the Z value corresponding to the probability of less than a given X. The NORMINV function returns the X value for a given probability, mean, and standard deviation.

FT6 *t* Value

Use the **TINV** function to calculate a critical value of the t distribution. Enter the function as **TINV(1 − *confidence level*, *degrees of freedom*)**. For a confidence level of 95%, enter **0.05** as the value for **1 − *confidence level***.

FT7 Critical Value of the Normal Distribution

Use the **NORMSINV** function to calculate a critical value of the normal distribution. Enter the function as **NORMSINV**($P<X$), where $P<X$ is the area under the curve that is less than X.

FT8 Cumulative Normal Probability

Use the **NORMSDIST** function to calculate the cumulative normal probability of less than the Z test statistic. Enter the function as **NORMSDIST**(Z **value**).

FT9 *t* Probability

Use the **TDIST** function to calculate a probability associated with the t distribution. Enter the function as **TDIST**(**ABS**(t), *degrees of freedom*, *tails*), where $ABS(t)$ is the absolute value of the t test statistic, and *tails* is either **1**, for a one-tail test, or **2**, for a two-tail test.

FT10 Chi-square Value

Use the **CHIINV** function to calculate a critical value of the chi-square distribution. Enter the function as **CHIINV**(*level of significance*, *degrees of freedom*).

FT11 Chi-square Probability

Use the **CHIDIST** function to calculate a probability associated with the chi-square distribution. Enter the function as **CHIDIST**(*critical value*, *degrees of freedom*), where *critical value* is the value of the chi-square test statistic.

ATT: Analysis ToolPak Tips (Microsoft Excel only)

The Analysis ToolPak add-in component of Microsoft Excel adds a number of statistical procedures. Use the instructions in section A.3 to verify that this add-in is installed on your computer. (The Analysis ToolPak is *not* usually installed when you initially set up Microsoft Excel and is not included with Excel 2008 for Macs.)

To use an Analysis ToolPak procedure, first select **Data Analysis** from the **Data** tab in Excel 2007 or select **Tools** then **Data Analysis** in earlier versions of Excel. In the Data Analysis dialog box that appears (see Figure D.1), click a procedure name and then click the **OK** button. For most procedures, a second dialog box appears, in which you make entries and selections to complete the procedure.

FIGURE D.1

Data
Analysis
dialog box

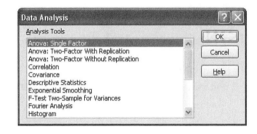

ATT1 Histogram Procedure

Begin by making sure that the worksheet that contains your data values also contains a column of bin values (see Tip CT5). In the Histogram dialog box, enter the cell ranges of the data values and bin values and check **Labels** if these ranges begin with a column heading. Check **Chart Output** if you want to create a histogram as well as the default frequency distribution.

ATT2 Descriptive Statistics Procedure

Place the data values for the variable to be summarized in a column, using the row 1 cells for column labels. In the Descriptive Statistics dialog box, enter that column range as the **Input Range**, click **Columns** in the **Grouped By** set, and check **Labels in First Row**. Then click **New Worksheet Ply** and check **Summary statistics**. A table of descriptive statistics appears on a new worksheet.

ATT3 t-Test: Two-Sample Assuming Equal Variances Procedure

Place the data for the two groups in separate columns, using the row 1 cells for column (group) labels. In the t-Test: Two-Sample Assuming Equal Variances dialog box, enter the group 1 cell range as the **Variable 1 Range** and enter the group 2 cell range as the **Variable 2 Range**. Enter 0 as the **Hypothesized Mean Difference** and check **Labels**. Enter 0.05 as the **Alpha** value, select the **New Worksheet Ply** option, and click **OK**. The results appear on a new worksheet.

ATT4 t-Test: Paired Two Sample for Means Procedure

Place the data for the two groups in separate columns, using the row 1 cells for column (group) labels. In the t-Test: Paired Two Sample for Means dialog box, enter the group 1 sample data cell range as the **Variable 1 Range** and enter the group 2 sample data cell range as the **Variable 2 Range**. Enter **0** as the **Hypothesized Mean Difference** and check **Labels**. Enter **0.05** as the **Alpha** value, select the **New Worksheet Ply** option, and click **OK**. The results appear on a new worksheet.

ATT5 ANOVA: Single Factor Procedure

Place the data of each group in its own column, using the row 1 cells for column (group) labels. In the ANOVA: Single Factor dialog box, enter the cell range for *all* of your data as the **Input Range**. Select **Columns** and check **Labels in First Row**. Enter **0.05** as the **Alpha** value, select the **New Worksheet Ply** option, and click **OK**. The results appear on a new worksheet.

ATT6 Regression Procedure

Place the data for each variable in its own column, using the row 1 cells for column (group) labels. Use the first column for the dependent variable Y and use the second and subsequent columns for your independent X variables. (Simple linear regression, discussed in Chapter 10, uses only one independent variable.) In the Regression dialog box, enter the cell range for dependent variable Y as the **Input Y Range** and enter the cell range for independent X variable or variables as the **Input X Range**. Check **Labels** and **Confidence Level** and enter **95** in the percentage box. To assist in a residual analysis, also check **Residuals** and **Residual Plots**. Click **OK**. The results appear on a new worksheet.

APPENDIX

E

Advanced Techniques

e.1 Using PivotTables to Create Two-Way Cross-Classification Tables

You use PivotTables (called DataPilots in OpenOffice.org Calc 3) to create two-way cross-classification tables from your data values. PivotTables are worksheet areas that act as if you had entered formulas to summarize data. PivotTables give you the ability to drill down, or look at, the unsummarized data values from which the summary information is derived.

You create a PivotTable by dragging variable names into a PivotTable form or template. The process varies depending on whether you use Excel 97–2003, Excel 2007 or later, or OpenOffice.org Calc 3.

For Excel 97–2003:

1. Open to the worksheet that contains your unsummarized data.

2. Select **Data** then **PivotTable Report** (Excel 97) or **PivotTable and PivotChart Report** (Excel 2000–2003).

3. In the Step 1 dialog box, click **Microsoft Excel list or database** as the source data and **PivotTable** as the report type. (You do not select a report type in Microsoft Excel 97; PivotTable is assumed.) Click **Next**.

4. In the Step 2 dialog box, enter the cell range of your unsummarized data. (The first row of this range must contain variable names.) Click **Next**.

5. In the Step 3 dialog box, click the **New worksheet** option as the location for your PivotTable. Then click **Layout**.

6. In the Layout dialog box, drag the label of the row variable to be summarized and drop it in the **ROW** area. Drag a second copy of this same label and drop it in the **DATA** area. (This second label changes to

Count of variable.) Drag the label of the column variable and drop it in the **COLUMN** area. (Figure E.1 shows a completed layout for the manufacturing plant study example in Section 2.1, "Presenting Categorical Variables.") Click **OK**.

7. Back in the Step 3 dialog box, optionally click **Options**, make changes to the formatting of the PivotTable, and click **OK** in that dialog box. (If you use Excel 97, the Layout dialog box appears as the Step 3 dialog box. When you click **Next** in this box, you then see a Step 4 dialog box that contains location options. No Options button exists in either the Step 3 or Step 4 dialog boxes in Excel 97.)

8. Still in the Step 3 dialog box, click **Finish** to create the two-way cross-classification table.

FIGURE E.1

Layout dialog box example

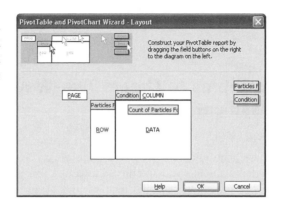

For Excel 2007 or later:

1. Open to the worksheet that contains your unsummarized data.

2. Select **Insert** then **PivotTable**.

3. In the Create PivotTable dialog box, leave the **Select a table or range** option selected and change, if necessary, the **Table/Range** cell range to the cell range of your unsummarized data. Then select the **New Worksheet** option and click **OK**.

4. In the **PivotTable Field List** task pane, drag the label of the row variable and drop it in the **Row Labels** box. Drag a second copy of this same label and drop it in the **Σ Values** box. (This second label changes to **Count of variable name**.) Drag the label of the column variable and drop it in the **Columns Labels** box. (Figure E.2 shows a completed layout for the manufacturing plant study example in Section 2.1, "Presenting Categorical Variables.")

5. Optionally, right-click the PivotTable and click **Table Options** in the shortcut menu that appears. In the PivotTable Options dialog box, adjust the formatting and display of the PivotTable and click **OK**.

FIGURE E.2

PivotTable
Field List
task pane
example

For OpenOffice.org Calc 3:

1. Open to the worksheet that contains your unsummarized data.

2. Select **Data**, then **DataPilot**, then **Start**.

3. In the Select Source dialog box, leave the **Current selection** option selected and click **OK**.

4. In the DataPilot dialog box, drag the label of the row variable and drop it in the **Row Fields** area. Drag a second copy of this same label and drop it in the Data Fields area. (This second label changes to **Sum - variable name**.) Double-click this second label and in the Data Field dialog box, select **Count** and click **OK**.

5. Still in the DataPilot dialog box, drag the label of the column variable and drop it in the **Columns Fields** area. Click **More** and select - **new sheet** - from the **Results to** drop-down list. Then click **OK**.

e.2 Using the FREQUENCY Function to Create Frequency Distributions

You can use the **FREQUENCY** function to create a frequency distribution table similar to the one created by the Analysis ToolPak **Histogram** procedure.

To use this function, first create a column of bins (see Tip CT5 in Appendix D for more information). Then, with your mouse, select the cells to contain the frequency counts (the cell range B4:B10 in the example below). Type, but do not press the **Enter** or **Tab** key, a formula in the form =FREQUENCY(*cell range of the data values, cell range of the bins*). Then, while holding down the **Control** and **Shift** keys (or the **Apple** key on a Mac), press the **Enter** key. This enters the formula as an "array formula" in all the cells you previously selected. To edit or clear the array formula, you must select all the cells, make your change, and again press **Enter** while holding down **Control** and **Shift** (or **Apple**).

The example below is based on the FREQUENCY worksheet found in the **Chapter 2 FREQUENCY** file. The worksheet includes formulas to calculate the total and percentage, if necessary. The *cell range of the data values* in the FREQUENCY function has been entered as Data!B2:B30 and not as B2:B30 because the data values are found on another worksheet—the Data worksheet. Using the name of a worksheet (followed by an exclamation point) as a prefix directs the spreadsheet program to that worksheet.

	A	B	C
1	Frequency Distribution of Fan Cost Index		
2			
3	Bins	Frequency	Percentage
4	199.99	=FREQUENCY(Data!B2:B30,A4:A10)	=B4/B$11
5	249.99	=FREQUENCY(Data!B2:B30,A4:A10)	=B5/B$11
6	299.99	=FREQUENCY(Data!B2:B30,A4:A10)	=B6/B$11
7	349.99	=FREQUENCY(Data!B2:B30,A4:A10)	=B7/B$11
8	399.99	=FREQUENCY(Data!B2:B30,A4:A10)	=B8/B$11
9	449.99	=FREQUENCY(Data!B2:B30,A4:A10)	=B9/B$11
10	499.99	=FREQUENCY(Data!B2:B30,A4:A10)	=B10/B$11
11	Total:	=SUM(B4:B10)	

E.3   Calculating Quartiles

Although Microsoft Excel and OpenOffice.org Calc 3 both contain a **QUARTILE** function, neither function calculates quartiles using the rules listed in Section 3.2. Instead, a series of formulas in **Chapter 3 Descriptive** calculates the first and third quartiles. Although a complete discussion of these formulas (see below) is beyond the scope of this book, the set of formulas uses the **IF** function to decide whether the calculations in steps 2 through 4 of the quartile rules of Section 3.2 are to be used. Judicial use of the **FLOOR, CEILING, ROUND**, and **SMALL** functions enable individual formulas to identify the ranks to be used as the data values that correspond to those ranks.

	F	G	H
14		Quartile Calculations	
15	First quartile rank	=(COUNT(A4:A13) + 1)/4	
16	Decimal fraction:	=IF(G15 = INT(G15), "#NA", G15 - INT(G15))	
17	= IF(G16 = "#NA", "use rank:", "use ranks:")	=IF(G16 <> "#NA", FLOOR(G15, 1), ROUND(G15, 0))	=IF(G16 <> "#NA", CEILING(G15, 1), "")
18	IF(G17 = "#NA", "use value:", "use values:")	=SMALL(A4:A13, G17)	=IF(G16 <> "#NA", SMALL(A4:A13, H17), "")
19	First Quartile:	=IF(G16 = "#NA", G18, (G18 * (1 - G16)) + (H18 * G16))	
20	Third quartile rank	=(3 * (COUNT(A4:A13) + 1))/4	
21	= IF(G20 = INT(G20), "", "Decimal fraction:")	=IF(G20 = INT(G20), "#NA", G20 - INT(G20))	
22	= IF(G21 = "#NA", "use rank:", "use ranks:")	=IF(G21 <> "#NA", FLOOR(G20, 1), ROUND(G20, 0))	=IF(G21 <> "#NA", CEILING(G20,1), "")
23	IF(G22 = "#NA", "use value:", "use values:")	=SMALL(A4:A13, G22)	=IF(G21 <> "#NA", SMALL(A4:A13, H22), "")
24	Third Quartile:	=IF(G21 = "#NA", G23, (G23 * (1 - G21)) + (H23 * G21))	

e.4 Using the LINEST Function to Calculate Regression Results

You can use the **LINEST** function to calculate regression results in OpenOffice.org Calc 3 or in Microsoft Excel versions starting with Excel 2003. These results are similar to those calculated by the Analysis ToolPak **Regression** procedure.

To use this function, first, with your mouse, select an empty cell range that is five rows deep and contains the number of columns that is equal to the number of your independent X variables plus one. For a simple linear regression model (see Chapter 10, "Simple Linear Regression"), select five rows by two columns; for a multiple regression model (see Chapter 11, "Multiple Regression") with two independent variables, select five rows by three columns.

Type, but do not press the **Enter** or **Tab** key, a formula in the form =LINEST(*cell range of the dependent variable*, *cell range of the independent variables*, **True**, **True**). Then, while holding down the **Control** and **Shift** keys (or the **Apple** key on a Mac), press the **Enter** key. This enters the formula as an "array formula" in all the cells you previously selected. To edit or clear the formula, you must select all the cells, make your change, and again press **Enter** while holding down **Control** and **Shift** (or **Apple**).

The results returned by the array formula are unlabeled. For a multiple regression model, some results appear as **#N/A**, which is not an error. Add labels in the columns immediately to the left and right of the results area to label the results, as shown in the example below.

The **SLR-LINEST** and the **MR-LINEST** worksheets of **Appendix E Regression** illustrate this labeling technique. Compare the labels to the corresponding ones in the **SLR-ToolPak** and **MR-ToolPak** worksheets, which are copies of the spreadsheet regression results shown in Chapters 10 and 11, respectively.

	A	B	C	D
1	Regression Analysis for Moving Company Study			
2				
3	Cubic Feet Moved Coefficient (b₁)	0.0501	-2.3697	Intercept Coefficent (b₀)
4	Cubic Feet Moved *Standard Error*	0.0030	2.0733	Intercept *Standard Error*
5	R Square	0.8892	5.0314	Standard Error
6	F	272.9864	34	Residual *df*
7	Regression *MS*	6910.7189	860.7186	Residual *MS*

APPENDIX

F

Documentation for Downloadable Files

This appendix lists and describes all of the files that you can download for free at **www.ftpress.com/youcanlearnstatistics2e** for use with this book.

F.1 Downloadable Data Files

Throughout this book, the file icon identifies downloadable data files that allow you to examine the data for selected problems.

Each data file is available as:

- An Excel .xls workbook file that can be opened by all Microsoft Excel versions and OpenOffice.org Calc 3
- Either a TI.83m matrix file or a TI .83l list file that can be loaded into any calculator in the TI-83 and TI-84 family

The following table identifies the columns of data found in each file and identifies the chapter or chapters in which the data is used. For the Excel workbook files, the columns of data map to the lettered columns of a worksheet. For the TI .83m matrix files, the columns map to the columns of the matrix variable [D]. In the special case of a single-column data file such as **Sushi** and **Times**, the single column of data is stored as a TI .83l list file.

Column names that appear in italics, such as *Day* in the **ADErrors** file, are row label columns and are not included in the TI matrix and list files.

AdErrors	*Day*, ads with errors, number of ads (Chapter 12)
Anscombe	Data sets A, B, C, and D—each with 11 pairs of *X* and *Y* values (Chapter 10)
Auto	Miles per gallon, horsepower, and weight for a sample of 50 car models (Chapter 11)
BankTime	*Day*, Customer *A* wait time, Customer *B* wait time, Customer *C* wait time, Customer *D* wait time (Chapter 12)
BB2008	*Team*, wins, earned run average (Chapter 10)
Cardiac	*Patient*, observer *A*, observer *B* (Chapter 8)
CatFood	Ounces eaten of kidney, shrimp, chicken liver, salmon, and beef (Chapter 9)
Cereals	Plant 1 cereal weights, Plant 2 cereal weights, Plant 3 cereal weights, and Plant 4 cereal weights (Chapter 9)
Chemical	*Batch* and viscosity (Chapters 2, 3, 5, and 6)
Cola	*Day*, total number of cans filled, and number of unacceptable cans (for a 22-day period) (Chapter 12)
Concrete	*Sample*, compressive strength after two days, and compressive strength after seven days (Chapter 8)
CostofLiving	*City*, apartment rent, and cost of a hamburger (Chapter 2)
GlenCove	Appraised value of house, land area in acres, and age of house (Chapter 11)
GolfBall	Design 1 distance traveled, Design 2 distance traveled, Design 3 distance traveled, Design 4 distance traveled (Chapter 9)
Hotels	*Year* and mean room rate in dollars (Chapter 2)
Intaglio	Surface hardness of untreated steel plates and surface hardness of treated steel plates (Chapter 8)
Math	Math scores using set A materials, math scores using set B materials, math scores using set C materials (Chapter 9)
MoviePrices	*Theatre chain* and cost of two tickets, a large popcorn, and two sodas (Chapters 3 and 6)
Moving	Labor hours, cubic feet moved, and number of large pieces of furniture moved (Chapters 2, 10, and 11)
Myeloma	*Patient*, measurement before transplant, and measurement after transplant (Chapter 8)
NBA2008	*Team*, number of wins, field goal (shots made) percentage, and field goal percentage allowed (Chapter 11)

NBACost	*Team*, ticket cost in dollars, and fan cost index (the cost of four tickets, two beers, four soft drinks, four hot dogs, two game programs, two caps, and the parking fee for one car) (Chapters 2 and 5)
Newspaper	*Day*, spot 1 blackness, spot 2 blackness, spot 3 blackness, spot 4 blackness, and spot 5 blackness. Also arranged as *Day* and newspaper blackness (.xls file only) (Chapter 12)
OrderTimePopulation	Population of order times from a website (Chapter 6)
Phone	Location 1 time (in minutes) to clear line problems and location 2 time to clear line problems for samples of 20 customer problems reported to each location (Chapter 8)
PropertyTaxes	*State* and property taxes per capita (Chapter 2)
Redwood	Height, diameter, and bark thickness (Chapter 11)
RestCost	Cost for a city restaurant meal, cost for a suburban restaurant meal (Chapters 3, 5, 6, 8)
Restaurants	*Location*, food rating, decor rating, service rating, summated rating, coded location (0 = urban, 1 = suburban), and price of restaurants (Chapter 10)
SamplesofOrderTimes	Sample 1 order times from a website, sample 2 order times, sample 3 order times....and sample 20 order times (Chapter 6)
Supermarket	*Pair*, sales using new package, sales using old package, and difference in sales (Chapter 8)
Sushi	Number of pieces needed to reach the maximum acceptable level of mercury (Chapters 3 and 6)
Times	Times to get ready (Chapter 3)
Transmit	*Day* and number of errors in transmission (Chapter 12)
Transport	*Day*, Patient *A* transport time (in minutes), Patient *B* transport time, Patient *C* transport time, and Patient *D* transport time. Also arranged as *Days* and patient transport times (.xls file only) (Chapter 12)
TVChannels	*Year* and the number of TV channels received (Chapter 2)
WholeFoods	*Item*, price at Whole Foods, and price at Fairway (Chapter 8)
Yield	New method 1 yield, new method 2 yield, and standard method yield (Chapter 9)

Use the TI DataEditor component of TI Connect (see Section A.C4) to transfer a downloaded TI data file to your calculator. Open TI Connect and click the TI DataEditor icon. When the DataEditor window opens, select **File →
Open** and navigate to the file you want to transfer. Then click the **Send File** icon to transfer the contents of the data file to your calculator.

calculator keys

Converting Matrix Variable Data into List Variable Data

Transferring one of the downloadable TI .83m matrix files to your calculator places new values in your calculator's matrix variable [D]. A matrix variable stores multiple columns, unlike a list variable which stores only a single column. For statistical functions that expect your data to be stored in a single-column list variable, you can extract a single column from matrix variable [D] and place it in one of the list variables in your calculator.

To extract a single column from matrix variable [D], press [2nd] [x^{-1}] [▶] to display the MATH menu and select 8:Matr→list(and press [ENTER]. In response to the Matr→list(prompt:

- Press [2nd] [x^{-1}] (to display the NAMES menu) and select 4: [D] and press [ENTER]. (This types the name of matrix variable [D].)

- Press [,] and then press the number key that corresponds to the number of the single column you want to extract. For example, if matrix variable [D] was storing the data of the **Auto** data file, you would press [2] to extract the horsepower data column.

- Press [,] [2nd] and then press the number key that corresponds to the list variable that you want to store the extracted data. For example, you would press [2] if you wanted to store the extracted data in list variable L2.

- Press [Enter].

You can also use the **Matrix-to-list** function to extract all of the columns of the matrix variable into a series of list variables in one operation. For example, if matrix variable [D] was storing the data of the **Auto** data file, you could extract the three data columns and place them in list variables L1, L2, and L3 by doing the following:

- Press [2nd] [x^{-1}] [▶] to display the MATH menu and select 8:Matr→list(and press [ENTER]. (This displays the Matr→list(prompt.)

- Press [2nd] [x $^{-1}$] (to display the NAMES menu) and select 4: [D] and press [ENTER].
- Press [,] [2nd] [1] [,] [2nd] [2] [,] [2nd] [3]. (This types the list ,L1,L2,L3)
- Press [Enter].

F.2 Downloadable *Spreadsheet Solution* Files

Also available for download are the Excel workbook files that are mentioned in the *Spreadsheet Solution* sections of this book. These workbook files are available in both the .xls format and the newer .xlsx (for Excel 2007 and later versions only). The following is a complete list of the Spreadsheet Solution Excel workbook files:

Chapter 2 Bar	Chapter 5 Poisson
Chapter 2 FREQUENCY	Chapter 5 Z Value
Chapter 2 Histogram	Chapter 6 Proportion
Chapter 2 Pareto	Chapter 6 Sigma Unknown
Chapter 2 Pie	Chapter 8 Pooled-Variance t ATP
Chapter 2 Scatter Plot	Chapter 8 Pooled-Variance t
Chapter 2 Time-Series	Chapter 8 Separate-Variance t
Chapter 2 Two-Way PivotTable	Chapter 8 Z Two Proportions
Chapter 2 Two-Way	Chapter 9 Chi-Square
Chapter 3 Descriptive ATP	Chapter 9 One-Way ANOVA ATP
Chapter 3 Descriptive	Chapter 10 Simple Linear Regression ATP
Chapter 3 Worked-out Problem 3	Chapter 11 Multiple Regression ATP
Chapter 5 Binomial	Appendix E Regression
Chapter 5 Normal	

Glossary

Alternative hypothesis (H_1)—The opposite of the null hypothesis (H_0).

Analysis of variance (ANOVA)—A statistical method that tests the effect of different factors on a variable of interest.

Bar chart—A chart containing rectangles ("bars") in which the length of each bar represents the count, amount, or percentage of responses in each category.

Binomial distribution—A distribution that finds the probability of a given number of successes for a given probability of success and sample size.

Box-and-whisker plot—Also known as a **boxplot**; a graphical representation of the five-number summary that consists of the smallest value, the first quartile (or 25th percentile), the median, the third quartile (or 75th percentile), and the largest value.

Categorical variable—The values of these variables are selected from an established list of categories.

Cell—Intersection of a row and a column in a two-way cross-classification table.

Chi-square (χ^2) distribution—Distribution used to test relationships in two-way cross-classification tables.

Coefficient of correlation—Measures the strength of the linear relationship between two variables.

Coefficient of determination—Measures the proportion of variation in the dependent variable Y that is explained by the independent variable X in the regression model.

Collectively exhaustive events—One in a set of these events must occur.

Common causes of variation—Represent the inherent variability that exists in the system.

Completely randomized design—Also known as **one-way** ANOVA; an experimental design in which only a single factor exists.

Confidence interval estimate—An estimate of the population parameter in the form of an interval with a lower and upper limit.

Continuous numerical variables—The values of these variables are measurements.

Control chart—A tool for distinguishing between common and special causes of variation.

Critical value—Divides the nonrejection region from the rejection region.

Degrees of freedom—The number of values that are free to vary.

Dependent variable—The variable to be predicted in a regression analysis.

Descriptive statistics—The branch of statistics that focuses on collecting, summarizing, and presenting a set of data.

Discrete numerical variables—The values of these variables are counts of things.

Error sum of squares (SSE)—Consists of variation that is due to factors other than the relationship between X and Y in a regression analysis.

Event—Each possible type of occurrence.

Expected frequency—Frequency expected in a particular cell if the null hypothesis is true.

Expected value—The mean of a probability distribution.

Experiments—A process that uses controlled conditions to study the effect on the variable of interest of varying the value(s) of another variable or variables.

Explanatory variable—The variable used to predict the dependent or response variable in a regression analysis.

F distribution—A distribution used for testing the ratio of two variances; also used in the Analysis of Variance and Regression.

First quartile Q_1—The value such that 25.0% of the values are smaller and 75.0% are larger.

Five-number summary—Consists of smallest value, Q_1, median, Q_3, largest value.

Frame—The list of all items in the population from which samples will be selected.

Frequency distribution—A table of grouped numerical data in which the names of each group are listed in the first column and the percentages in each group are listed in the second column.

Histogram—A special bar chart for grouped numerical data in which the frequencies or percentages in each group are represented as individual bars.

Hypothesis testing—Methods used to make inferences about the hypothesized values of population parameters using sample statistics.

Independent events—Events in which the occurrence of one event in no way affects the probability of the second event.

Independent variable—The variable used to predict the dependent or response variable in a regression analysis.

Inferential statistics—The branch of statistics that analyzes sample data to reach conclusions about a population.

Joint event—An outcome that satisfies two or more criteria.

Level of significance—Probability of committing a Type I error.

Mean—The balance point in a set of data that is calculated by summing the observed numerical values in a set of data and then dividing by the number of values involved.

Mean squares—The variances in an Analysis of Variance table.

Median—The middle value in a set of data that has been ordered from the lowest to highest value.

Mode—The value in a set of data that appears most frequently.

Multiple regression—Regression analysis when there is more than one independent variable.

Mutually exclusive events—Events are mutually exclusive if both events cannot occur at the same time.

Normal distribution—The normal distribution is defined by its mean (μ) and standard deviation (σ) and is bell shaped.

Normal probability plot—A graphical device to evaluate whether a set of data follows a normal distribution.

Null hypothesis (H_0)—A statement about a parameter equal to a specific value, or the statement that no difference exists between the parameters for two or more populations.

Numerical variables—The values of these variables involve a count or measurement.

Observed frequency—Actual tally in a particular cell of a cross-classification table.

p chart—Used to study a process that involves the proportion of items that have a characteristic of interest.

p-value—The probability of computing a test statistic equal to or more extreme than the result found from the sample data, given that the null hypothesis H_0 is true.

Paired samples—Items are matched according to some characteristic and the differences between the matched values are analyzed.

Parameter—A measure that describes a characteristic of a population.

Pareto chart—A special type of bar chart in which the count, amount, or percentage of responses of each category are presented in descending order left to right, along with a superimposed plotted line that represents a running cumulative percentage.

Percentage distribution—A table of grouped numerical data in which the names of each group are listed in the first column and the percentages in each group are listed in the second column.

Pie chart—A circle chart in which wedge-shaped areas ("pie slices") represent the count, amount, or percentage of each category and the circle (the "pie") itself represents the total.

Placebo—A substance that has no medical effect.

Poisson distribution—A distribution to find the probability of the number of occurrences in an area of opportunity.

Population—All the members of a group about which you want to draw a conclusion.

Power of a statistical test—The probability of rejecting the null hypothesis when it is false and should be rejected.

Probability—The numerical value representing the chance, likelihood, or possibility a particular event will occur.

Probability distribution for a discrete random variable—A listing of all possible distinct outcomes and the probability that each will occur.

Probability sampling—A sampling process that takes into consideration the chance that each item will be selected.

Published sources—Data available in print or in electronic form, including data found on Internet websites.

Range—The difference between the largest and smallest values in a set of data.

Region of rejection—Consists of the values of the test statistic that are unlikely to occur if the null hypothesis is true.

Regression coefficients—The Y intercept and slope terms in the regression model.

Regression sum of squares (SSR)—Consists of variation that is due to the relationship between X and Y.

Residual—The difference between the observed and predicted values of the dependent variable for given values of the X variable(s).

Response variable—The variable to be predicted in a regression analysis.

Sample—The part of the population selected for analysis.

Sampling—The process by which members of a population are selected for a sample.

Sampling distribution—The distribution of a sample statistic (such as the mean) for all possible samples of a given size n.

Sampling error—Variation of the sample statistic from sample to sample.

Sampling with replacement—A sampling method in which each selected item is returned to the frame from which it was selected so that it has the same probability of being selected again.

Sampling without replacement—A sampling method in which each selected item is not returned to the frame from which it was selected. Using this technique, an item can be selected only once.

Scatter plot—A chart that plots the values of two variables for each response. In a scatter plot, the X axis (the horizontal axis) always represents units of one variable and the Y axis (the vertical axis) always represents units of the second variable.

Simple linear regression—A statistical technique that uses a single numerical independent variable X to predict the numerical dependent variable Y and assumes a linear or straight-line relationship between X and Y.

Simple random sampling—The probability sampling process in which every individual or item from a population has the same chance of selection as every other individual or item.

Six Sigma—A method for breaking processes into a series of steps in order to eliminate defects and produce near-perfect results.

Skewness—A skewed distribution is not symmetric. An excess of extreme values are in either the lower portion of the distribution or the upper portion of the distribution.

Slope—The change in Y per unit change in X.

Special causes of variation—Represent large fluctuations or patterns in the data that are not inherent to a process.

Standard deviation—Measure of variation around the mean of a set of data.

Standard error of the estimate—The standard deviation around the line of regression.

Statistic—A numerical measure that describes a characteristic of a sample.

Statistics—The branch of mathematics that consists of methods of processing and analyzing data to better support rational decision-making processes.

Sum of squares among groups (SSA)—The sum of the squared differences between the sample mean of each group and the mean of all the values, weighted by the sample size in each group.

Sum of squares total (SST)—Represents the sum of the squared differences between each individual value and the mean of all the values.

Sum of squares within groups (SSW)—Measures the difference between each value and the mean of its own group and sums the squares of these differences over all groups.

Summary table—A two-column table in which the names of the categories are listed in the first column and the count, amount, or percentage of responses are listed in a second column.

Survey—A data collection method that uses questionnaires or other approaches to gather responses from a set of participants.

Symmetry—Distribution in which each half of a distribution is a mirror image of the other half of the distribution.

t **distribution**—A distribution used to develop a confidence interval estimate of the mean of a population and to test hypotheses about means and slopes.

Test statistic—The statistic used to determine whether to reject the null hypothesis.

Third quartile Q—The value such that 75.0% of the values are smaller and 25.0% are larger.

Time series plot—A chart in which each point represents a response at a specific time. In a time series plot, the X axis (the horizontal axis) always represents units of time and the Y axis (the vertical axis) always represents units of the numerical responses.

Two-way cross classification table—A table that presents the count or percentage of joint responses to two categorical variables (a mutually exclusive pairing, or cross-classifying, of categories from each variable). The categories of one variable form the rows of the table, whereas the categories of the other variable form the columns.

Type I error—Occurs if the null hypothesis H_0 is rejected when it is true and should not be rejected. The probability of a Type I error occurring is α.

Type II error—Occurs if the null hypothesis H_0 is not rejected when it is false and should be rejected. The probability of a Type II error occurring is β.

Variable—A characteristic of an item or an individual that will be analyzed using statistics.

Variance—The square of the standard deviation.

Variation—The amount of dispersion, or "spread," in the data.

Y intercept—The value of Y when $X = 0$.

Z score—The difference between the value and the mean, divided by the standard deviation.

Index

A

α, 145

alternative hypothesis, 142-143, 359

analysis of variance (ANOVA). *see one-way analysis of variance*

ANOVA summary table, 189

arithmetic mean. *see mean*

arithmetic and algebra review, 301–310

attribute control charts, 266

B

β, 145

bar chart, 20–21, 359

binomial distribution, 90–91, 359

box-and-whisker plot, 58–62, 359

C

calculator keys,
    binomial distribution, 93
    box-and-whisker plot, 62
    Chi-square tests, 193
    confidence interval estimate for the mean (σ unknown), 130
    confidence interval estimate for the proportion, 133
    converting matrix variable data into list variable data, 356–357
    entering data, 9–10
    initial state, 294
    keystroke conventions, 293

mean, 54

median, 54

menus, 296

multiple regression, 252–253

normal probabilities, 105

normal probability plot, 107

one-way analysis of variance (ANOVA), 193

Poisson probabilities, 97

pooled-variance t test for the difference in two means, 162

residual analysis in multiple regression, 254

simple linear regression, 228

standard deviation, 54

technical configuration, 297

TI Connect, 298–299

variance, 54

Z test for the difference in two proportions, 156

categorical variable, 3, 359

cell, 359

central limit theorem, 121

certain event, 73

Chi-square distribution, 182, 360

Chi-square distribution tables, 320–321

Chi-square test, 179

classical approach to probability, 77

cluster sampling, 9

coefficient of correlation, 221, 360

coefficient of determination, 220, 360

coefficient of multiple determination, 248–249

collectively exhaustive events, 73–74, 360

common causes of variation, 269, 360

complement, 74

completely randomized design. *see one-way analysis of variance*

confidence interval estimate, 125–127, 360
    for the mean (σ unknown), 127–130
    for the proportion, 131–133
    for the slope, 226–227, 252

continuous numerical variables, 360

continuous values, 3

control chart factor tables, 338

control charts, 268–270, 360
    p-chart, 271–276
    range (*R*) chart, 279–281
    $\bar{X}$ chart, 279–283

control limits, 270

critical value, 144, 360

D

degrees of freedom, 129, 188, 360

dependent variable, 208, 360

descriptive statistics, 5, 360

discrete numerical variables, 360

discrete values, 3

discrete probability distribution, 84

DMAIC model, 267–268

double-blind study, 6

downloadable files, 353–357

E

elementary event, 72

empirical approach to probability, 77

equation blackboard,
    binomial distribution, 91–92
    Chi-square tests, 191–193
    confidence interval estimate for the mean (σ unknown), 129
    confidence interval estimate for the proportion, 132
    confidence interval estimate for the slope, 227
    mean, 45
    mean and standard deviation of a discrete probability distribution, 89

median, 47

one-way analysis of variance (ANOVA), 191–193

p-chart, 274–276

paired *t* test, 169–170

Poisson distribution, 96

pooled-variance *t* test for the difference in two means, 164–165

quartiles, 49

range (*R*) chart, 281

range, 52

regression measures of variation, 218–220

slope, 214–217

standard deviation, 55

standard error of the estimate, 222

t test for the slope, 225–226

variance, 55

chart, 282

Y intercept, 217

Z scores, 56

Z test for the difference in two proportions, 158–159

event, 71, 360

expected frequency, 180, 360

expected value of a random variable, 85–86, 360

experimental error, 187

experiments, 6, 360

explanatory variable. *see independent variable*

F

F distribution 188, 360

F distribution tables, 322–337

F test statistic, 189

factor, 186

five number summary, 59, 361

frame, 7, 361

frequency distribution, 26–28, 361

H

histogram, 28–29, 361

hypothesis testing 141, 361

hypothesis testing steps, 147

I

independent events, 76, 361

independent variable, 208, 360–361

inferential statistics, 5, 119, 361

J–K
joint event, 72, 361

L
least-squares method, 210
left-skewed, 57
level of significance, 361

M
mean, 43–45, 361
mean squares, 361
  among groups (*MSA*), 188
  total (*MST*), 188
  within groups (*MSW*), 188
measures of
  central tendency, 43–51
  position, 47–51
  variation, 51–56
median, 44, 46–47, 361
misusing charts, 32–33
mode, 47, 361
multiple regression model, 245, 361
mutually exclusive, 74, 361

N
net regression coefficients, 247–248
normal distribution, 98–104, 361
normal distribution tables, 312–315
normal probability plot, 105–106, 361
null event, 73
null hypothesis, 142, 362
numerical variable, 3, 362

O
observed frequency, 362
one-tail test, 148
one-way analysis of variance, 186–196, 359
  assumptions, 195
operational definition, 4
overall *F* test, 249–250

P
p-chart, 271–276, 362
p-value, 147, 362
paired *t* test, 166–171
parameter, 4, 362
Pareto chart, 22–23, 362
percentage distribution, 26–28, 362
pie chart, 21–22, 362
PivotTables, 347–349
placebo, 6, 362

point estimate, 123
Poisson distribution, 94–95, 362
pooled-variance *t* test, 160–166
population, 2, 362
power of the test, 146, 362
practical significance, 145
primary data sources, 6
probability, 73, 362
  rules, 74
probability distribution for discrete
  random variables, 83–84, 362
probability sampling, 8, 362
published sources, 6, 363

Q
quartiles, 47–51

R
random variable, 72
range, 51–52, 363
range (*R*) chart, 279–281
red bead experiment, 276–278
region of nonrejection, 144
region of rejection, 144, 362
regression model prediction, 213
regression analysis, 207
residual analysis, 363
  simple linear regression, 223–224
  multiple regression, 250–251
regression assumptions, 222
response variable. *see dependent variable*
right-skewed, 57

S
sample, 2, 363
sampling, 8, 363
sampling distribution, 119–120, 363
  of the mean, 120–122
  of the proportion, 123
sampling error, 124–125, 363
sampling with replacement, 9, 363
sampling without replacement, 9, 363
scatter plot, 30–31, 363
secondary data sources, 6
shape, 57–58
simple linear regression, 208, 363
  assumptions, 222
simple random sampling, 8–9, 363
Six Sigma, 267–268, 363
skewness, 57, 364
slope, 209–210, 364

special causes of variation, 269, 364
spreadsheet operating conventions, 299
spreadsheet solutions,
  bar and pie charts, 22
  binomial probabilities, 93
  Chi-square tests, 193
  confidence interval estimate for the
    mean (σ unknown), 130
  confidence interval estimate for the
    proportion, 133
  entering data, 12
  frequency distributions and
    histograms, 29
  measures of central tendency and
    position, 50
  measures of variation, 54
  multiple regression, 254
  normal probabilities, 104
  one-way analysis of variance
    (ANOVA), 193
  paired t test, 168
  Pareto charts, 24
  Poisson probabilities, 97
  pooled-variance t test for the
    difference in two means, 163
  scatter plots, 32
  simple linear regression, 227
  two-way tables, 26
  Z test for the difference in two
    proportions, 158
spreadsheet technical configurations,
  299–300
spreadsheet tips,
  Analysis ToolPak tips, 343–345
  Chart tips, 339–341
  Function tips, 341–343, 349–351
standard deviation, 52–55, 364
standard deviation of a random
  variable, 86–87
standard error of the estimate, 221–222,
  364
standard (Z) scores, 55–56
statistic, 4, 364
statistics, 1, 364
stratified sampling, 9
subjective approach to probability, 78
sum of squares,
  error (SSE), 218–219, 360
  regression (SSR), 217–219, 363
  total (SST), 187, 218–219, 364

  among groups (SSA), 187, 364
  within groups (SSW), 188, 364
summary table, 19–20, 364
surveys, 7, 364
symmetric, 57, 364

T–U
t distribution, 364
t distribution tables, 316–319
tables of the,
  Chi-square distribution, 320–321
  control chart factors, 338
  F distribution, 322–337
  normal distribution, 312–315
  t distribution, 316–319
test of hypothesis,
  Chi-square test, 179–186
  for the difference between two
    proportions, 153–159
  for the difference between the means
    of two independent groups,
    160–166
  for the slope, 227–228
  in multiple regression, 251–252
  one-way analysis of variance,
    187–193
  paired t test, 166–171
test statistic, 143–144, 364
time–series plot, 29–30, 364
total quality management, 265–267
treatment effect, 187
two-tail test, 148
two-way cross-classification tables,
  24–26, 365
Type I error, 145, 365
Type II error, 145–146, 365

V–W
variable, 3, 365
variable control charts, 268
variance, 52–55, 365

X
$\bar{X}$ chart, 279–283

Y
Y intercept, 209, 365

Z
Z scores, 55–56, 365